ADVANCE PRAISE *for* QUA

MW00659191

"Agriculture is at a crossroads, and we are beginning to realize that the current system is not providing the stability in our production system needed to overcome the variations due to soil degradation and weather during the growing season. This book offers valuable insight into a path forward to enhance our soils for generations to come and to provide the quantity and quality of food we need for food security. This book will require you to think about changes needed in our agricultural systems."

—Jerry Hatfield

"If you're like me, you like to connect with folks that make you think. This book does that. I've listened to most of the podcasts, and I enjoy John's innate ability to connect folks from many different facets of research, bringing them to a combined platform. This book allows you to go deeper, with the resources and data in the background as supporting evidence. I've read it from start to finish, and I'm looking forward to future volumes to expand my personal knowledge base."

—Loran Steinlage

"No matter what form of production agriculture you are involved with, *Quality Agriculture* is a must read. John uses his expertise to delve deep into agoecology with many of the industry's pioneers."

—Gabe Brown

"Knowledge about regenerative farming systems comes in many forms from many farmers, researchers, ag educators, and consultants. The challenge is sorting through all the information, learning what is possible, and putting together a working system for the land that matters. These interviews certainly expand and clarify the knowledge base of professionals who have spent their careers studying regenerative agriculture. A great read—you won't be disappointed."

—Gary F. Zimmer

Advance Praise *for* Quality Agriculture

Inspiring and insightful interviews with true leaders in regenerative agriculture.
—David R. Montgomery, author of *Growing A Revolution: Bringing Our Soil Back to Life*

John Kempf, who has one of the most fertile minds in agriculture today, has done us a huge favor with this book by multiplying his knowledge with like-minded experts he's interviewed recently. Collectively, this disruptive innovation of how we grow food will be the foundation of how we farm in the future!
—Steve Groff, farmer, Cover Crop Coach, and author of *The Future-Proof Farm: Changing Mindsets in a Changing World*

John Kempf's book on regenerative agriculture is a great collection of the wisdom and experience of twelve different agricultural pioneers from around the country. John directs the discussion, unlocking one door after another on fascinating insights each of these experts have discovered in their work. I have met several of the participants over the years but learned much more from these interviews than from my casual visiting with them and hearing them speak. In every case, what the reader finds is the destructive nature of modern, industrialized agriculture that is focused on artificial inputs and unsustainable outputs. Each interviewee directs our attention back to the complex interrelationships found in the soils, plants, and microbiology of nature. This is a must read for everyone who farms and everyone who eats.
—Robert Quinn

The human mind is a world to itself. This book is a collage of splendid agriculture minds with a common theme: embrace and emulate the power of life. The collective wisdom of this book will help you glean the kernels of truth in a research world built on a wrong premise—asking the wrong questions. If you garden, farm, or just enjoy the ecological world, build the correct foundation in your mind by reading this book.
—Ray Archuleta

COPYRIGHT © 2020 JOHN KEMPF
Regenerative Agriculture Publishing
All rights reserved.

No part of this publication may be reproduced, stored in a retrieval system, or transmitted in any form or by any means, electronic, mechanical, photocopying, recording, scanning or otherwise, except as permitted under Section 107 or 108 of the 1976 United States Copyright Act with the prior written permission of the author.

QUALITY AGRICULTURE

Conversations about Regenerative Agronomy with Innovative Scientists and Growers

ISBN 978-1-7348445-0-4 *paperback*
978-1-7348445-1-1 *ebook*

Quality Agriculture

Conversations *about* Regenerative Agronomy *with* Innovative Scientists *and* Growers

John Kempf

CONTENTS

Introduction

MAINSTREAM AGRONOMY is based on a series of fallacies. That plants only absorb nutrients in the form of soluble ions. That mites, nematodes, insects, fungi, bacteria, and viruses attack plants indiscriminately. That pesticides are effective long-term solutions. All of these commonly held ideas are false.

Present-day mainstream agronomy and plant nutrition is based on the disproven idea that soil chemistry balance is the only parameter worth measuring and managing. This has led to farmers buying fertilizers their crops don't need, thereby supplying excessive and imbalanced nutrition. This results in increased pest susceptibility and increased need for pesticide applications to prevent diseases and insects from consuming the now unhealthy crops.

Because of this chemistry-based approach to crop production—with little consideration for soil biology—farmers are led to purchase inputs they don't need with money they often don't have because their advisers claim they need them to remain competitive. Combined with a singular emphasis on high yields as the only meaningful metric of success, with little thought of crop quality, this approach often results in food and feed crops of questionable nutritional integrity.

Due to this mindset, our soil's natural fertility has been greatly reduced. Weeds have become more difficult to control. Pest pressure has increased. More pesticide applications are required. And the nutritional quality of many foods has dropped.

Many growers with a few decades of experience know this to be true because they have observed it on their own operations. They have observed how pest pressure and crop performance have changed. They are certain there has to be a different way—a better way—but they are unclear where to begin.

For many growers, the mainstream approach is all they have ever learned; it's all they know. But they also know that now is the time to change. We now have the knowledge and information to make better choices.

This book is intended to showcase the in-depth knowledge and wisdom that is available to growers today—information that wasn't as easily available even a decade ago. It is a collection of interviews that I conducted with some of the most knowledgeable and influential voices in regenerative agriculture today. As the host of the *Regenerative Agriculture Podcast* I have been privileged to ask the questions I'm personally curious about, knowing that many others share my desire to learn.

I have been asked many times to put together written resources from the work I've done over the years as a consultant. Providing the words of my mentors and colleagues in this field seemed like the best way to start doing this.

The voices in this book discuss the large and growing body of knowledge about agroecology. Many of them, including Tom Dykstra, Michael McNeill, and Jerry Hatfield, discuss how to manage nutrition to produce disease- and insect-resistant crops. Gabe Brown and Kris Nichols talk about the negative effects of chemical fertilizers and how biological diversity—both below the surface and above—are key for healthy farms. Gary Zimmer, Robert Kremer, and Jonathan Lundgren consider how to greatly reduce or eliminate the need for pesticides and how to increase crop yields while simultaneously increasing quality, regenerating soil health, and reducing fertilizer inputs. Gerald Pollack shares how structured water rapidly transports nutrients and molecules through plants. Chili grower Ed Curry discusses his passion for open-pollinated seed breeding. Matt Kleinhenz describes why nutritional integrity and density matters. And the legendary Don Huber delves deep into the important disease-suppressive effects of the crop prior to a cash crop.

The knowledge of each of these pioneers is invaluable. Their ideas will surely be adopted on a large scale around the globe as the economic benefits of regenerative agriculture become realized.

More than even its economic benefits, regenerative growing methods are vital because they return joy to our work. Farming becomes fun again when we work with natural systems instead of working against them. When we experience the benefits of working with biological systems and robustly healthy plants, it quickly becomes obvious that we need less fertilizer. Once we have the assurance of personal experience that our crops are not susceptible, we don't need to worry about possible diseases or insects. And of course, when input costs are reduced while yield and quality are increased, the improved profitability provides a foundation to make all of farming fun again. Farming can only be a successful lifestyle if it is first a successful business. The subjects of this book reiterate these revelations over and over again.

As an agronomy adviser working with the Advancing Eco Agriculture (AEA) team, I have been privileged to help develop regenerative management systems on hundreds of thousands of acres and to observe the large untapped potential that exists on many farms. I am passionate about developing regenerative agriculture systems that improve soil health, produce crops that are completely resistant to diseases and insects, and produce food of such an exceptional quality that we can have a legitimate conversation about growing food as medicine.

I have been blessed to have exceptional mentors. I've discovered that there are many people with incredible knowledge and information about soil and plant systems, and about how to develop regenerative agricultural systems. However, much of this knowledge is scattered and not widely known. It's found all over the place. A little bit of it has been published in peer-reviewed publications, but there are many incredible stories and a lot of knowledge that has not been published or shared with many people.

I started AEA in 2006 to bring this knowledge together in a more coherent fashion, incorporating it into products and growing systems that growers can easily put into practice. It's my personal mission to have these regenerative agricultural systems become adopted globally and become the mainstream—the status quo against which all other growing systems are compared. To help achieve this goal, I want to share the knowledge that we have learned over the last fourteen years and make it available to everyone.

While AEA has developed products that embody the principles of regenerative agriculture systems and made them easier for growers to apply, this knowledge and these principles can be applied anywhere. When they are applied properly, they will always increase farm profitability and resilience to climate stress.

These interviews are inspirational to me, and I'm confident they will be to you as well. My hope is that this book will give you a very different perspective on how to think about agronomy and crop production. I'm confident it will provide many ideas that you can easily implement on your operation. When you find the information in this book valuable, please share it with others, leave a review, and let others know what you find useful.

The views and content of my parts of these interviews—including any unintentional errors—are mine alone, and the same holds for the subjects. Each conversation has been edited for clarity and for length.

Acknowledgments

MANY THANKS to the following people, who have inspired me and who have made this book—and all my work—possible.

To my parents, Emanuel and Mary Kempf—for being innovators, for providing an example of doing things differently, and for giving their family permission to challenge the way things have always been done. I am very grateful for your support.

To the mentors who have guided me on this pathway: Bruce Tainio, Michael McNeill, Don Huber, Jerry Hatfield, and the many others who have taught me through their writings.

To each of my interviewees and guests, who so graciously shared their knowledge and wisdom.

To the team at Advancing Eco Agriculture, who I work every day to make our vision of regenerative agriculture systems a reality, along with the growers we are privileged to partner with. Our collective experience keeps us grounded and on the cutting edge.

To my editor, Paul Meyer, who transformed the rough interview transcripts into smooth reading.

Developing Disease Resistance *and* Regenerating Soil Health

MICHAEL MCNEILL

I've had the fortune to see that old kind of farming, and I see, perhaps, a need to return to a portion of what my grandfather was doing back in those days. I don't necessarily mean that we've got to sell our tractors and buy horses, but some of the things he did to improve soil health, and to maintain soil fertility, are things that we really need to be thinking about today.

Some tillage is not as bad as the herbicides, not as bad as anhydrous ammonia, and not as bad as the high-salt fertilizers. They tend to be more of an issue. When you put them all together, it overwhelms the soil-life system.

Micronutrients are extremely important to plant growth. They are readily and easily chelated by the pesticides that we use. And once you tie them up, you start shutting down significant pathways.

Disease and insects are Mother Nature's garbage collectors—getting rid of the bad stuff, the weak plants.

I believe that the soil can grow an extremely healthy, high-yielding plant with minimal additions of inputs. There are plenty of minerals in the soil if you treat it properly.

Stop poisoning the soil.

MICHAEL MCNEILL is one of the wise elders of agriculture. Born and raised in Iowa, he has witnessed the transition into conventional chemical agriculture, its wide-scale adoption and subsequent problems, and more recently, the advent of more regenerative growing methods.

Michael has a very impressive background in agronomy, with degrees in plant physiology and statistical genetics. He earned his PhD from Iowa State University and worked for several years as a plant pathologist for the U.S. Army. He has had a very long and successful career in the Midwest, serving as a plant breeder and research station director.

Michael founded Ag Advisory Ltd. in 1983, and today, he continues to work as a crop consultant, helping growers face a broad range of different challenges.

JOHN: Michael, can you begin by telling us a little bit about your background and the work that you are doing on a day-to-day basis on your operation?

MICHAEL: I was born and raised on a farm northwest of Algona, Iowa, and I'm still farming there today. I've watched it go from horse-drawn equipment to our modern-day tractors that are GPS-guided across the field. It's been quite a transition that I've gotten to see, and I've learned a lot along the way. I've had the fortune to see that old kind of farming, and I see, perhaps, a need to return to a portion of what my grandfather was doing back in those days.

I don't necessarily mean that we've got to sell our tractors and buy horses, but some of the things he did to improve soil health, and to maintain soil fertility, are things that we really need to be thinking about today. In the past couple years, I've decided to go totally organic on our farm, which makes it a bit unusual—but somewhat commonplace for our community. There are about forty-two thousand acres of organic crops in the two-county area here where I live.

JOHN: Wow, that's grown quite substantially.

MICHAEL: It's a huge change. There is a sort of a stigma when you talk about organic farmers. Many times, folks think of the guy that's farming the twenty- or forty-acre patch that's covered with weeds. But the farmers in our community that we're working with range all the way from a half section—320 acres of land—to 15,000 acres. Those farms are pretty much weed free. They're excellent farmers. They've learned a lot about weed management, weed control, soil fertility, and soil health. I'm pretty excited to see that change. But it does remind me a bit of the things that my grandfather taught me years ago—the way he took care of his crops.

JOHN: Can you tell us a little bit about your professional work—what you do professionally in addition to the farming operation?

MICHAEL: WELL, I left the farm for a few years and got my formal education at Iowa State University, where I majored in agronomy. I got my BS degree in soils and soil fertility. And then I became quite interested in how soil impacted plant growth. I have my master's degree in plant physiology. Then I was intrigued with how genetics work and how the environment impacted genetics, so I ended up getting my PhD in quantitative genetics.

Along that same path, I also became interested in plant diseases and got a minor in plant pathology. After leaving the university, I spent some time at a facility in Frederick, Maryland, where we studied the impact of diseases that could be used as weapons and how we could defend our crops against that type of thing. I learned a great deal about how fertility and genetics can create an environment that can allow us to protect ourselves from any pathogenic invasion.

JOHN: That sounds intriguing. Where did that lead you?

MICHAEL: I became a corn breeder, and I worked for a major company for twelve years managing a corn-breeder research station. I started their soybean breeding project, as well as continuing in the corn-breeding project. During that time, farmers would ask me more fertility and soil fertility questions than they would genetic questions. I kept trying to explain how the two work together.

That was about the time that GMOs were being invented. We were using gene guns and blasting gold dust through genes and trying to move them from one organism into another. We started taking genes out of different species and moving them around—taking genes out of bacteria and putting them in a corn plant. This really didn't sit well with my thoughts about how our species was created and why. Who is man to mess with that? We were opening Pandora's box to who knows what when we started doing this.

I chose to step out of that type of breeding work—sticking mainly with my quantitative genetics approach versus the gene approach. I chose to work more with soil health, soil fertility, and genotype by fertility interactions. That led me into a profession of agricultural consulting. I started working with farmers and explained how all this worked, and it got to be a pretty exciting adventure. I started with a farmer—he had 160 acres—and then his neighbor added another 160 to it, and I'm now up to about 165,000 acres that I work on. That's sort of a history of how I got to where I am today.

JOHN: When you say you have about 165,000 acres that you work on today, what is the scope of the work you do on each of these farms?

MICHAEL: Most of it is working with soil health and soil fertility, and helping growers select the right genetics for the fertility programs that they're working with. Soil health is becoming a bigger and bigger issue for me to deal with. When I first started, it wasn't a really big issue. It's huge now. I'm devoting more of my time now to soil health than I ever thought I would.

JOHN: I'd love to talk about that a little bit—when you say that soil health didn't used to be a big issue, and now you're spending a lot of time on it, what changed with soil health? How are you managing it differently today than you were twenty or thirty years ago?

MICHAEL: Well, it's interesting that you would ask me that, John. The other day, I was cleaning out a drawer in my desk, and I found some old pictures that I had taken back in 1972 or 1973 of crops that were growing. I had some close-ups and some overviews of the field. The thing that I noticed was how healthy the plants were. There were no disease lesions on them anywhere. The corn plants were just perfect. The whole field was that way.

It's really hard to find a field today that is that way. I was looking at the weeds that were growing along the fence rows, and they were big and healthy and looked great. They don't look so good today, comparatively speaking. And you say, "Well, maybe that's a good thing!" No, it's not. The whole area that we're farming is unhealthy. It makes me ask the question, What's changed?

To me, the big difference from that era until today is that farmers have been drawn into big ag: You need to use herbicides. You don't want to use a cultivator. You have to farm more land. So you use herbicides, but herbicides are doing things to the soil because they're all chelators. Now the plants become a little bit imbalanced in the nutrition that they're taking up, and you find more disease—you find more insect pressure. So you start using fungicides and insecticides—more chelators, more poisons being dumped onto the ground. You're pretty impressed with how they work. The field is perfectly clean, and weed free—excellent. The diseases were dramatically reduced. The fungicides worked really well. The corn borers and some other of the insects that were issues went away. It was magic. The chemistry was totally magic—it looked beautiful.

But as time went on, the chemistry started poisoning the good things that were in the soil. Today, I'm called out to look at farms where the guy's production has dropped off dramatically and where the soil is virtually dead.

JOHN: When you say the production has dropped off dramatically, what have you observed?

MICHAEL: Looking at ten-year crop insurance records, the guy was getting 190 to 210 bushels per acre and had around a 200-bushel, 10-year average. Excellent, excellent yields. Now it's getting 70- and 80-bushel yields. That's dramatic, and it will put him out of business very quickly.

JOHN: That is very dramatic.

MICHAEL: This isn't just happening on a little field here, a farm there. I'm seeing 8,000- and 10,000-acre farms that this has happened to. That really, really woke me up. I started seeing this about five years ago. I've been working with these growers who are asking me whether I can help them remediate that. Can I help bring the farm back? In a three- to four-year period, we've had pretty good success. I would say we're back now at where we were when this crashed.

The farmers are excited that they can now take it to a different level—to the 250-bushel range or greater. They can see growth and potential and do what they're doing. They've moved away from GMO crops, and they've particularly moved away from glyphosate.

JOHN: These are very important pieces for us to talk about and to understand, Michael. When you have a degraded system like that—where there are suppressed yields and suppressed soil health, as you're describing it—how do you go from depressed yields of 70 to 90 bushels per acre back up to 200, with aspirations of going back up to 250 bushels per acre? How do you achieve that?

MICHAEL: It's a long, hard task. There aren't any silver bullets. You have to figure out what was going wrong and stop doing that—that's number one. Number two, you're going to have to look at what it's going to take to remediate the soil. Has the soil become really hard—hard like a road? I get penetrometer readings where it takes five hundred pounds of downward pressure to penetrate the top two inches of the soil—that's hard. That's just like a gravel road. A crop will not grow in that.

When they till it, it's breaking up into chunks. Then when it rains, it puddles and it just seals over. We get no oxygen into the soil. You have to incorporate some tillage, and then you have to start providing some food for the microbial life—which is almost nonexistent. It's not nonexistent because you can bring it back—that's the good news. If you don't let this thing go too long, you can bring it back.

Now whether we're bringing all of it back or not, I don't know. But once you get it started coming back, then you can look at inoculating with mycorrhizae and some of the things—the pseudomonads, the actinomycetes—that could be missing and stimulate them. But first you have to get oxygen into the soil, get the water working correctly, and get the food right. There's no magic in inoculating the soil—if it's loaded with poison, it will kill your inoculant. You have to fix that problem first before you try inoculating. You wouldn't have to do an inoculation, but it does speed it up—you gain about a year, maybe two years, when you do that.

I see people thinking they're buying a magic silver bullet by inoculating, but then they continue to do the things that caused their soil to die in the first place. They're not winning. They're losing.

JOHN: You've iterated several times that you have to stop doing what inflicted the damage in the first place. What I heard you saying was that it's really the herbicides and fungicides and the insecticide applications that are causing this degradation of soil health. I heard you mention that these herbicides and these various pesticides that people are applying are actually chelation agents.

Why do you believe that these products are the causal agent for the suppression of soil health? Couldn't it also be the extensive tillage that we had for a number of decades and some of these other contributing factors?

MICHAEL: I have some farms that I feel are way overtilled. They're organic farmers. They really do till excessively, in my mind. But it doesn't seem to be bothering the soil at all. It isn't quite as good as I'd like to see it, but as long as they're keeping their organic matter up—preventing erosion, using cover crops, and that sort of thing—the tillage in itself doesn't seem to be doing as much damage as I originally thought it would.

Having said that, you have to be careful which tillage tools you use. A disc is not a very good tillage tool to be using—it causes compaction; it fractures the soil structure much worse than a tined implement that you could pull through—whether that be a V-ripper or a narrow-pointed field cultivator. These kinds of things do not seem to do the structural damage that I see with things like the disc, or even like a moldboard plow or a field cultivator with sweeps on it.

JOHN: In essence, you're saying that tillage doesn't have the damaging effects on soil health that the herbicides do, from your perspective.

MICHAEL: It's not as bad as the herbicides, not as bad as anhydrous ammonia, and not as bad as the high-salt fertilizers. They tend to be more of an issue. When you put them all together, it overwhelms the soil-life system.

JOHN: I understand the impact of anhydrous ammonia and salt fertilizers—both of those are very oxidizing and can have the potential to produce a lot of damage to the soil's microbial community. But I don't understand how herbicides would have that same effect. You mentioned herbicides being chelating agents. From your perspective, how is it that herbicides and these various pesticides have such a damaging effect on soil health?

MICHAEL: We have not paid a lot of attention to micronutrients in the soil. Micronutrients are extremely important to plant growth. They are readily and easily chelated by the pesticides that we use. And once you tie them up, you start shutting down significant pathways. That's where my physiology training and background came into play—when I started seeing a lot of these physiological processes being shut down.

An example people are probably familiar with is that if you chelate manganese and tie it up, you shut down the shikimate pathway. When you shut that down, diseases can move in very quickly because that's sort of the plant's immune system, if you will. If you shut that down, you have to buy fungicides. You put on the fungicides to protect your plant from the disease that's invaded, and then you start killing more of the fungal life in the soil. It's a vicious, vicious cycle that you've set up.

JOHN: You mentioned that you believe that perhaps we should go back to some of the practices you learned from your grandfather. What are the things you learned from him that you think would be relevant and appropriate today?

MICHAEL: One of the things was that he was very good at crop rotation. We've gotten ourselves into what I'd almost call it a monoculture—a "biculture," if you will—where we grow corn and soybeans, or soybeans and corn. It's just not working well with the soil microbial life.

Back in the day, he was using horses, and he needed feed for those horses. So he raised oats. Oats have a real good way of cleansing the soil—making more silica available for plant growth and taking toxins out of the soil. So he was keeping the soil clean, just because he needed to produce fuel for his power.

We don't need that today, so a lot of guys ask me, "Why should I grow oats? I don't need to grow oats." And I say, "Well, think about that as an excellent cover crop." So plant oats; get them growing. The winter will kill oat plants. You don't have to use a herbicide to kill it in the spring, and you've got all these good benefits—you've retained nutrition in the soil; you've retained nitrogen. You've gotten a very fibrous root system that is going to improve the soil structure. Plenty of food is there for microbial life. You're going to have some suppression of weeds in the spring with that oat cover. We can still use that crop today—we can rotate—but think about using it as a cover crop.

He also grew clovers and alfalfa and other legumes, which of course benefited the nitrogen for the corn production. Can we do that? Yes, we can. But we need to change the way we look at our land. We need livestock in the cropping areas. We've taken them out because you can grow huge numbers of cattle in feedlots, and they grow better in a warmer climate. That's true. But then you're shipping all of those nutrients into that area, and it becomes more of a waste product than a fertilizer. There is no better fertilizer than animal manures.

JOHN: Michael, we've been circling around this topic of soil health and the impacts of tillage, herbicides, animal manures, cover crops, and so forth. At the beginning, you mentioned that there's a correlation between soil health and the diseases that are present. You mentioned some work you were doing in Maryland—studying diseases as a weapon. What did you learn from that experience? And how does all of that tie into what we're talking about?

MICHAEL: Let's say you want to use a fungal disease as a weapon—that you can get this disease introduced into the soil. Not only does it kill a crop this year—it'll continue to kill it into future years. So hey, that's a pretty good weapon—you shut down a people's food supply. They have a problem.

If you have good pseudomonas bacteria in the soil, they act as a policeman in the soil, if you will, and they'll take out the pathogenic fungi that can arise. But if you use products like glyphosate—that's an antibiotic type of product—you're going to kill all the pseudomonas, and then you have no protection. It's very easy to get a huge population of *Fusarium* going in the soil, which probably is a pathogenic *Fusarium*—or *Pythium* or phytophthora. You've lost the natural balance. If you have that balance, though, the pathogenic fungi are not going to do much to you. Your good bacteria will clean it right up.

JOHN: So you can actually have a disease-suppressive soil where you don't have challenges with those pathogenic fungi. I think I also heard you mention that you were working on developing solutions to those diseases as weapons. What were the types of solutions that you were working on?

MICHAEL: There are all kinds of approaches. If you need a fast cure, of course, you've got to look at chemistry and the fungicides and that sort of thing. But what you find is that if you get the soil contaminated, how do you fix it? If you put anything on it, you're going to kill everything in the soil. Using a soil sterilizer is not necessarily a great idea. But there is microbial life in the soil that will hold everything in balance. If you have the right nutrition available, everything will take care of itself.

I'll use you as an analogy, John. If your nutrition gets pretty poor, you're going to get pretty run down, and you're going to be very susceptible to all kinds of diseases. Would you agree?

JOHN: Oh, I think that's just the story of the people who are trying to sell me supplements! [I'm being sarcastic—I take many supplements, and I believe they are important.]

MICHAEL: THAT'S funny—but when that occurs, you can take this supplement or this drug to prevent the disease, but you're still improperly nourished. You're going to get another disease, and then you're going to get another disease. But if you get your nutrition back and properly balanced, and everything is at the correct level, your immune system starts to function properly. A good share of your immune system is in your digestive tract—there are a lot of microbes working for you.

The soil is no different. You get those microbes working for you; you're going to stay healthy. The soil is going to stay healthy, and so are the plants.

JOHN: Are you saying that when you manage the nutritional balance of the soil and the microbial population of the soil, that it's possible to grow crops that don't have disease?

MICHAEL: Yes. When a plant is perfectly healthy, it's very hard to get a disease to invade it, and an insect will not even stop to look at it. Why is that? It's because an unhealthy plant cannot convert the sugars it's produced into complex sugars—starches and lignin—which insects and diseases can't use. They can use simple sugars and the nitrate nitrogen in the plant. The nitrate nitrogen is taken up by the plant, and it's immediately converted into amino acids and proteins in a healthy plant. An unhealthy plant—a plant that does not have the right mineral balance to make all those processes and cycles work—will have a pretty heavy load of nitrate in it—a fantastic food for the insect. They can detect that, and they will land on that plant and feed on it. Disease and insects are Mother Nature's garbage collectors—getting rid of the bad stuff, the weak plants.

JOHN: These are things that we have talked about and that I've spoken about quite extensively. We developed this diagram called the plant health pyramid, where we describe how plants become resistant to different types of diseases and different types of insect pests at different stages of physiological development. This is something that we've observed in the field very clearly. It's really exciting and a lot of fun when it starts functioning the way it was designed to function. All of a sudden, disease and insect pressure dramatically decreases. In many cases, you can say that it completely disappears. That's really exciting. That's a lot of fun.

MICHAEL: It really is. That takes me way back to the early years—when I looked at plants and they were not having any of these issues. The micronutrient levels were excellent. All the processes were functioning properly. And there was very little disease or insect pressure.

JOHN: It's very interesting to hear your perspective—of how things have changed over a forty- or fifty-year time period.

MICHAEL: I didn't really realize how much had changed until I pulled out those photos the other day.

JOHN: You don't suppose those photos are just one-offs from a really exceptional year?

MICHAEL: No. They were not taken for that reason. Some of those photos were pictures of livestock that were going to be sold; they were standing in the field or by a fence. I was looking beyond them and looking at the crop. If it was a photo that was taken of an exceptionally good crop or something, I would say that, yes, you have a point. But no, these photos were taken of and focused on other things—the crop was in the background. But as I observed it, it was like, "Check that out"—it was really looking good.

JOHN: Michael, what is something that you've puzzled over for a really long time? What's really caught your attention in the agriculture space that you've been working on?

MICHAEL: Something that I've finally figured out, I think, was the impact of the lack of availability of micronutrients in our crops. I was doing tissue testing, for example, and I had adequate copper and iron and manganese and magnesium and calcium—everything looked good. What I didn't realize was that a lot of those minerals were chelated. They were tied up into a form that the plant could not use—yet they showed up on a chemistry test when we tested the tissue. When I finally figured that out, then everything started to gel for me.

JOHN: I think what you're saying is that these various minerals and trace minerals were being chelated inside the plant tissue by the herbicides and fungicides that growers were applying.

MICHAEL: Yes. When I tested the plant, it had adequate levels. But when I looked at the plant, it was showing deficiency symptoms. You could look at it and just tell that there was a zinc deficiency or a manganese deficiency; it was obvious. But when I tested it, it was fine. Why was that? It's when I learned about this chelation issue and how it can be such a problem.

JOHN: This is something we've been monitoring for a number of years. And it seems that, in some cases, sap analysis reports those a bit more accurately. And perhaps that doesn't take all the chelation into account—but of course it's still extracting nutrients that are held within the plant sap, and it's still possible for them to be chelated. We do see the sap analysis correlate more accurately to what the plants are actually showing visually.

MICHAEL: I would agree. I think the sap analysis has been a good step forward. When I figured out this chelation effect, that's when it really gelled for me and I could understand why I was seeing deficiency symptoms in what, on paper, looked to be an appropriately healthy plant.

JOHN: We've been talking about a number of different things that you seem to have a very different framework on. What are some things you believe to be true that many other people don't believe to be true?

MICHAEL: I believe that the soil can grow an extremely healthy, high-yielding plant with minimal additions of inputs. There are plenty of minerals in the soil if you treat it properly. I think most people would disagree with me on that. They have "proven" that you have to put on fertilizers to get good yields.

JOHN: Wow.

MICHAEL: I'm going to stick with that because I've proven it to myself—that you can do that.

JOHN: You're saying that you can actually grow healthy, high-yielding crops without adding fertilizers?

MICHAEL: Yes.

JOHN: How do you do that? How does that work?

MICHAEL: That's a complex question with a very complex answer. The key is creating healthy soil that allows plant roots to go deep into the soil, to extract the minerals they need. There's a vast sea of minerals that are available. If you're starting to see suppression of crop yield, and then you add fertilizer, and the yield comes back—that's because you're only using the top few inches of the soil. The roots are not healthy enough to penetrate deeper—to actually mine the minerals that are there. Our soil is nothing but minerals.

JOHN: This is something that I've talked about as well. The question that I often get is, "Aren't you going to deplete the soil of minerals if you're not adding fertilizers?"

MICHAEL: My quick reply to that is, "Try and take all the salt out of the ocean." You can deplete minerals in a rooting zone—I'll grant you that. But what you need to do is expand your ability to search in a bigger rooting zone. And you need to add mycorrhizae into that equation because you want a really big rooting zone. Let the mycorrhizae work for you.

JOHN: I think this is a much bigger topic that we perhaps need to unwrap at some point in the future. It's very exciting—the idea that we can grow really healthy crops without constantly adding inputs.

When we talk about developing regenerative agriculture ecosystems today, there's a lot of discussion about sustainability. But we can't have a legitimate conversation about sustainability and a sustainable agriculture as long as we are constantly importing such large volumes of inputs. We first need to have a regenerative model—where we regenerate soil health and plant health to a much higher plateau of performance. Only at that point can we have a true conversation about sustainability. This is really exciting, because what you were describing is the foundation of what I would describe as true sustainability.

Here's the counterpoint to the question I asked: What do you not believe to be true that many other people do believe to be true?

MICHAEL: Most people believe that you have to put a lot of inputs in, in order to get a decent yield. They're not understanding how the system works. The way the system will work is not what we're trying to do today. We're trying to swim upstream, and it's just not working.

JOHN: I would agree with you that we're trying to force things to move in a direction that they don't naturally want to flow. We can talk about the science behind developing exceptionally healthy crops, and we can make it sound really complicated. But at the end of the day, the reality is that it was designed to work. It did work very successfully for a very long period of time. If we can give it the framework and the management that it needs to really flow well, then it can work very effectively. You just need to allow that to happen.

MICHAEL: Yes, I agree.

JOHN: What technology or ideas are you really excited about? What are you excited about in the future of agriculture?

MICHAEL: One of the big things is food production. We've done a lot of work with our "corn-soybean game," if you will. We've tried to make foods out of those crops. But these foods are not necessarily very healthy for us. I think that if we started concentrating on food production, then agriculture has a fantastic opportunity. We're wasting some of the best soils on earth to produce fuel for our vehicles. We're really missing the point.

JOHN: Just a moment ago, you mentioned the word "opportunity." Where do you see the big opportunities in agriculture today?

MICHAEL: I see the most success right now in growers who are looking to improve the sustainability of their farming operations. When they do that, they're finding that they rely on fewer and fewer inputs. They're getting higher-quality food and feed production at a lower cost, and they are profitable—in a time when agriculture is struggling in terms of profitability. But those growers are not struggling at all.

JOHN: What is the one book that you recommend most often to growers?

MICHAEL: A book that I really like is *Mineral Nutrition and Plant Disease*. It was edited by Lawrence Datnoff, Wade Elmer, and Don Huber. I find it to be a really useful book in helping you understand how minerals affect plants.

When you get a feel for that and for how that all works, then all these things that you and I have been talking about all of a sudden start making a lot of sense. That's a pretty good book to start with.

JOHN: In terms of agronomic sciences, I would say it's in the top two or three that I personally reference most often. So it definitely is a very valuable resource.

MICHAEL: THAT's a pretty decent one to start with. Then, once you have that under your belt, you can start delving into some of the books that talk about the microbial life in the soil and what it takes to improve that. Because once you're doing that, everything else falls into place for you.

JOHN: Yes, the plants are certainly the photosynthetic engine that drives soil biology. There's a really great book that was published back in the seventies, if I'm not mistaken, by the Russian Academy of Sciences, entitled *Soil Microorganisms and Higher Plants*, by N. A. Krasil'nikov.[1] It's incredibly powerful—incredibly insightful reading. It's a fairly in-depth book, but it's amazing—both how much we had already learned forty years ago and how much we have failed to put into practice. There's a lot of knowledge there that we haven't applied consistently in commercial-scale production agriculture, for sure.

Michael, what is the one action that you would advise all growers to take right now that could make the biggest difference in their operations?

MICHAEL: Stop poisoning the soil.

JOHN: I guess that's easy!

MICHAEL: It's real simple—just stop.

JOHN: That sounds simple. It sounds easy to do—but how? How do you manage that?

---

1 "Soil Microorganisms and Higher Plants," Soil and Health Library, 2020, https://soilandhealth.org/book/soil-microorganisms-and-higher-plants/.

MICHAEL: It is a challenge if you've spent most of your life doing things one way. Stopping doing something is not necessarily easy. But that's the one action growers need to take to be successful.

JOHN: Are there transition steps that can be taken to move away from that? What are some of your growers who have moved away from using herbicides doing?

MICHAEL: I have seen a full array of actions—from taking baby steps to jumping off the cliff—100 percent stop. I have seen growers—from the smaller, 300- to 400-acre growers to the 10,000- to 15,000-acre growers—step off the cliff. It's worked really well for them. I was really concerned about some of the larger growers, but I found that they had the management ability and the resources to make it happen. Once they understood what they were doing and why they were doing it, they were very successful.

I think that's something that most people don't believe. I get that thrown in my face almost every day. "I can't do that—I have too big an operation." I really enjoy throwing it back—"Well I know somebody who has." Those successful large growers have not necessarily added more hired men or anything. The one thing that they have added—if they've made a mistake or a failure—they've had to employ a large number of people for a short period of time to hand-weed a field. If they made a mistake, that's the only fix there is.

JOHN: I'm struggling with this a little bit myself as well. When you use the words "stepping off a cliff," are you talking about eliminating 100 percent of all herbicide applications right out of the gate? What does that mean, exactly?

MICHAEL: All pesticide applications.

JOHN: Aren't you going to lose your crop to potential disease and insect pests when you do that?

MICHAEL: When you do that, you'd better have read that book that I just suggested—so that you understand that you need to have the right micronutrient balance to keep that plant healthy enough to protect itself. You can do that through starter fertilizers, foliar feedings—multiple foliar feedings—you can pull it off.

JOHN: What are some of the failures of growers who have tried to do this, and what has been their degree of success?

MICHAEL: By and large, I have had all successes. I'm trying to think of a failure, but I really can't think of any. I make sure they really understand and know what they're doing when they do it. I've had a few where they missed a field or two, timing-wise—a rain caught them, and they didn't get the weeds taken care of when they should have. But they were able to get it cleaned up—to the point where it did not suppress yield.

JOHN: Wow. How do their yields compare?

MICHAEL: I think that their yields have been going up. That's what's been somewhat shocking. I want to be sure it's attributed to that—not just necessarily a good growing season. Because we've had some good growing seasons recently. But their yields have continued to climb quite rapidly. They've moved to a different yield plateau.

JOHN: So you're saying that their yields are actually higher now than they were when they were using herbicides and pesticides regularly?

MICHAEL: Yes.

JOHN: Well, that's exciting, because those are the same types of things that we've observed in the fruit- and vegetable-production world. And those are really the types of regenerative systems that we seek to create and to establish, and I absolutely agree with you that those are possible.

From a management perspective, the one piece we often do a bit differently on fruit and vegetable crops we work on is that we don't usually advise people to "step off the cliff," to borrow your terminology. Rather, we advise growers to manage nutrition and to regenerate soil health to a higher plateau of performance—to the point where growers earn the right to eliminate pesticides. Then, all of a sudden, we don't have problems with powdery mildew anymore. We don't have problems with spider mites anymore. We don't have problems with leafhoppers anymore. When we get to that much higher plateau, and we no longer have the problems, then we start cutting and eliminating pesticide applications.

It seems a bit scary to me—when you're managing a crop that is really valuable—to suggest eliminating all pesticide applications immediately. But obviously, you've been successful in doing so.

MICHAEL: Yes, it's worked. It was really scary when I first started doing that. But I've learned the few things that you have to be sure to accomplish: getting the soil as healthy as you can and helping the plants be as healthy as you can. That's pretty hard to do when stepping off the cliff. But it can be done.

Suppressing Disease *for* Future Crops

DON HUBER

Manganese has a very dynamic relationship with the soil, and also with many of the fungi. There are organisms—mycorrhizae—that increase the uptake of manganese, as well as zinc and phosphorus and some of the other nutrients. If they're not functional, you miss that ability to absorb and to interact with a tremendous volume of the soil.

When we're farming, we're really managing an ecology. It's not a matter of a silver bullet for this problem or a stinger missile for another. It's really a matter of having ecology work for us and support the plant. If we don't do that and we upset the system, then we compromise the overall quality and productivity potential we have in our soil.

Crop rotation probably has the most dynamic effect on the soil microbial population. We can do that primarily through crop sequencing. The crop that immediately precedes your target crop is going to provide about 85 percent of the overall disease-suppressing effect. We can accomplish that with cover cropping.

By learning what your soil is—what your biology is—you can control most diseases through management—unless you're doing something that is very detrimental or has long-term effects on biology. If you increase carbon dioxide just a small amount, it'll have an exponential increase in photosynthesis.

It's critical that we recognize the stewardship we have when it comes to the soil—and our own survivability—rather than just accepting the changes and saying, "Well, there's nothing we can do about it now." There's a tremendous amount we can do. It's a matter of understanding that we're managing an ecology.

HOW MUCH OF AN EFFECT DOES CROP ROTATION REALLY HAVE? According to Dr. Don Huber, 85 percent of disease suppression is a result of the crop that was planted prior to a cash crop. Don shares this and many other fascinating insights in this interview.

Don is Professor Emeritus of Plant Pathology at Purdue University. After receiving his PhD at Michigan State University, Don worked for several years for the army, retiring in 1995 as a Colonel after forty-one years of active and reserve service. He joined the faculty of Purdue in 1971. He has authored or coauthored hundreds of papers, presentations, and other publications, including coediting *Mineral Nutrition and Plant Disease*, one of the key texts on plant health.

Don has had an extraordinary life. He has contributed a great deal to our collective knowledge of plant pathology, developing disease-suppressive soils, and the possible long-term impacts of herbicides and pesticides.

JOHN: Don, you've had some very rich life experiences. What are some of the memorable moments that have led you to where you are today?

DON: I just have fun wherever I am. At any particular time, you have a few challenges, but those help you grow and help you focus on what you really want to do—or how you think you can be of greater service.

JOHN: What was something in your research work that you puzzled over for a really long time—that you struggled to figure out?

DON: There have been a number of them. When you study soil ecology, you're looking at a black box. The pieces to the puzzle come a little slower than they might for some other programs or other areas, but you start working on the border, and then you start filling in the center. As that puzzle builds, you get a little better picture of all the dynamics that are involved. I think that the greatest challenge is to gain some comprehension of what that black box really is doing—its dynamic nature and its critical importance and the overall production scheme.

JOHN: What has been something that has surprised you?

DON: How everything can be working at one time but seem to be so contradictory.

JOHN: Can you give me an example of that?

DON: A lot of your nutrient relationships—where you have microorganisms that are responsible for changing the valence states of various minerals so that they're more available or less available. You have those going on in both directions at the same time in some capacities. It's an issue with manganese or iron or some of those things. Some of the secondary functions come into play so that all of that can take place and manifest in a very positive manner. Even though what you might be looking at—or the tests that you have—may not show the complete picture, you have to realize that it has to be going on in order to complete the cycle.

It's a matter of either developing the techniques or understanding how all of those organisms interact—the ecological niches that make the system work. Everything isn't just one big pool with somebody stirring the whole thing around. You really have a community of functions that are taking place at the same time, but you don't have the same gas station at every corner or a grocery store at every corner. You have each one of those different functions taking place in its own little scheme of things. The overall system is a very functional and very dynamic relationship relative to the plan. It's neat.

JOHN: One of the pieces that you and I have discussed in the past is the challenge that we are seeing today with manganese availability. I would say that as much as 80 percent or more—perhaps even 90 percent or more of the crops that we work with today—come back showing inadequate levels of manganese. What are the major factors that contribute to that?

DON: Manganese has a very dynamic relationship with the soil, and also with many of the fungi. There are organisms—mycorrhizae—that increase the uptake of manganese, as well as zinc and phosphorus and some of the other nutrients. If they're not functional, you miss that ability to absorb and to interact with a tremendous volume of the soil—where that mineral might be in short supply.

The other thing is that you have bacteria that are responsible for the valence state. You have the oxidizing groups. You have the reducing groups. The plant can utilize only the reduced form of manganese—the Mn^{2+} form. Mn^{4+} form is nonavailable, but we see it primarily in the soils that have high phosphate levels or high oxidative relationships—the manganese can be there and yet not be available for uptake. We see it with many of our pathogens because the pathogens utilize manganese oxidation as a virulence factor.

We looked at several thousand isolettes of *Gaeumannomyces graminis*, which causes take-all all over the world. We evaluated those, and we found that there was one characteristic that was common in all virulent farms—manganese oxidation. If the wheat had oxidized manganese, it would never resist the disease. The same thing for rice blast. The same thing for isolettes of *Streptomyces scabies* and a number of other pathogens. The ability to oxidize manganese to a nonavailable form—and to compromise the resistance of the plant to those pathogens—if the pathogens can't bring about that compromising of the availability of manganese by converting it to an oxidized form, the fungus is essentially just a good saprophyte in the soil.

Same thing with many bacteria. We see these direct effects on mineral availability being involved not just in growth and quality and nutrient density but also in susceptibility or resistance to disease. You have the virulence relationship of the pathogens with bacteria and fungi in the soil, and that's related to those minerals that are necessary for the plant's defenses. Those minerals are also directly related to the growth and resistance of the plants to those pathogens in their overall physiological function. It all fits together very nicely if the system is balanced—if it's favorable.

That's one of the things that we can adapt to. When we're farming, we're really managing an ecology. It's not a matter of a silver bullet for this problem or a stinger missile for another. It's really a matter of having ecology work for us and support the plant. If we don't do that and we upset the system, then we compromise the overall quality and productivity potential we have in our soil.

JOHN: You said that there are a number of pathogens that are dependent on manganese oxidation. If they're unable to oxidize manganese, they just become saprophytes in the soil profile. Are they dependent on that manganese oxidation directly—do they individually require it? Or are they just producing a manganese-deficient plant that is now susceptible to invasion?

DON: Both of those statements would be correct. They don't necessarily need the oxidation. Some of them are also reducing organisms. In other words, if you change the environment—or if you change the association that they have with other organisms—then they may be strong-reducing versus strong-oxidizing organisms.

We see that especially with the *Pseudomonads* and a number of other organisms—you change the soil environment, and they can benefit you, or they can be synergistic, or they can even be a direct pathogen, involved in compromising that resistance. The microorganisms use those minerals just like a plant does, or just like we do. Our metallonutrients or strong transition elements or electron transfer and physiological processes are the cofactors for enzyme function. We don't require very much of them, but if you don't have that specific cofactor that's involved for an enzyme, that enzyme isn't going to do any work for you—it's just another protein that's sitting there. About 80 percent of our proteins in plants are what we call metalloproteins, where the metallo part is a cofactor. It's a small part but a very critical one, as far as function of that physiological pathway.

JOHN: In essence, what you're describing is that as long as plants have adequate availability of reduced manganese, they have resistance to all the diseases that you described.

DON: It would be very, very critical for that resistance—for the physiological functions in in the plant—without those minerals.

JOHN: One of the pieces that you and I have discussed in the past is the populations of some strains of *Verticillium* in the soil profile—when you have a crop that is infected versus a crop that is not infected. How do populations change when you have a crop infection for some of these pathogens?

DON: The plant is providing nutrients and resources for the pathogen. *Verticillium* is a common soil-inhabiting organism, not just a soil-colonizing organism. It's soil inhabiting—it's able to survive in the soil on its own. It really doesn't need the plant to continue its life cycle in the soil. A soil colonizer is an organism that can have a very high population as long as there is a host there to provide it a nutrient base.

Most *Penicillium* and *Aspergillus* species will isolate in very high populations, but they're only associated with high organic matter and those materials that provide a rich nutrient base. If you take the organic matter out of the picture—the nutrient base—they aren't able to survive very well.

Compare that with *Mucor* or *Fusarium* or *Rhizoctonia* or *Phytophthora* or regular soil-borne pathogens, and you'll find that these organisms are going to be in the soil at a similar population regardless of whether there is a nutrient basis—whether the plant's there or not. The population of *Fusarium* may bounce around when you add the plant because you're providing an additional nutrient source for it, but its population is high enough in almost all situations to cause extreme disease even in the absence of the host plant. You can have 100 percent infection if you inoculate sterile soil, for instance.

The difference is where you have a lot of other organisms that can suppress that virulence relationship that the pathogen has, which is maintained either in a dormant state or with some of the fungi. We refer to that as fungistasis—a dormant state—because you have an intimate association with bacteria or other organisms in association with that fungus that prevent pathogenicity. Those other organisms may be utilizing the extracellular enzymes that would oxidize manganese or perform other functions that reduce its virulence, but they don't necessarily reduce the population of the pathogen to a low-enough level to cause extreme disease. It's just that they change its ability to cause disease—so that it's more a saprophyte than it is a strong pathogen.

JOHN: That's absolutely fascinating—that we can have these soil-inhabiting pathogens that are present in large populations. Obviously, these populations can shift, but it's intriguing to me that essentially what you're describing is that an infection is not the result of the presence of a pathogen. It's the result of the level of plant health and a result of the microbial ecology in the rhizosphere.

DON: That's correct. The soil-inhabiting organisms are dependent on a host for nutrient base and for a disease relationship. With the soil-inhabiting organisms, we see a dramatic effect of crop rotation—a link between susceptible crops. We see that relationship with the soil inhabiting not because of presence or absence of the pathogen but by the specific interrelationships we have with different organisms with the pathogen.

With the soil colonizers—organisms that can colonize as long as there's a nutrient source for it—those have more direct effects of population dynamics and the typical epidemiological programs or processes going on in the soil. With the actual soil colonizers, you see the interaction of other organisms and disease suppression.

Some examples of soil inhabitants are *Fusarium*, *Rhizoctonia*, or *Verticillium*. Organisms that would respond to population dynamics—nematodes, or things like anthracnose or other organisms—once the nutrient source—the host residue—is decomposed and gone, the pathogen isn't able to survive in the soil against the tremendous biological buffering and the interactions that take place in a normal soil.

JOHN: What are the tools that growers have available to them to manage soil ecology most effectively and to develop a disease-suppressive soil profile? What are the most effective management strategies for managing soil ecology?

DON: From a dynamic standpoint, crop rotation has the greatest effect. It's going to depend on a particular pathogen that's involved. Each crop manages or influences a certain a group of organisms in the soil, and those then are going to enter into that relationship.

We found with alfalfa—when it was introduced as a crop—was a total failure for a few years, until we recognized the importance of symbiotic nitrogen fixation. We started adding legume inoculants in order to have that free nitrogen. That made it an economic crop for us. Without that, we were having to add nitrogen and other minerals, but with that symbiotic relationship, we get nitrogen for nothing, essentially, by having those organisms fix atmospheric nitrogen into amino acids for the plant to utilize.

That would be an example of that symbiotic relationship from the nutrient standpoint. But crop rotation will influence each group of organisms. It makes it very easy, if we're developing a desert soil, to build up disease suppressiveness. We can do that strictly by the crop sequence. We can build a biological buffering capacity in the soil in a desert condition; those soils have very low microbial populations and activity due to the lack of water. When you get a cultivated soil, it becomes much more difficult to shift that biology in the direction that you want. For some pathogens, we've never been able to find a crop rotation that will do that. But for most of them, we can modify the disease expression—even independently of the population of the pathogen—by favoring certain other groups of organisms in that relationship.

Crop rotation probably has the most dynamic effect on the soil microbial population. We can do that primarily through crop sequencing. The crop that immediately precedes your target crop is going to provide about 85 percent of the overall disease-suppressing effect. We can accomplish that with cover cropping. You don't necessarily have to eliminate your economic crop—cover cropping can do much of that work while providing additional synergy from a nutrient standpoint.

When you consider how to manage that soil, we're looking at the influence of root exudates and the influence of a particular crop—what it leaves in the soil and how it influences nutrients that then are available to the pathogen or the other saprophytic organisms that influence soil quality.

Certainly, tillage has as a very dynamic effect. Biomobility is very important, especially in the temperate climates. Time of tillage becomes very important because you're disrupting an ecology that's established in a no-till situation, and you're trying to activate or suppress different organisms that have established a king-of-the-hill relationship with the system that you have. Tillage really changes that overall ecology by getting some air into the soil, enabling nutrient distribution and stirring up those organisms. Some of them are kind of like we are—if they don't have a burr under their saddle, they don't do much work for us. You're stirring them up.

In a temperate environment, where we have winter seasons and summer seasons, you'll find that fall tillage, after your crop, will have a more pronounced or a more dynamic relationship than you'll see with spring tillage. In fall tillage, you open up the soil; it's very oxidative for organisms to have that fresh charge of breath. And then over winter, you reestablish those ecological niches so that your anaerobic functions then take place. You still have your aerobic functions taking place.

Reduction of nutrient forms, or oxidation of nutrient forms, are then rebalanced in the spring, when we do a lot of tilling. You don't have time during that crop that immediately follows the tillage operation to reestablish a lot of those ecological niches; they take some time to develop. You get a different effect. It can be a negative effect, or it may be a more positive effect. It depends on the crop and the overall disease organisms or overall nutrient status in a particular soil—whether it's clay loam or silt loam or sandy loam. These ecological relationships shift in a dynamic relationship relative to time.

Moisture becomes a very important factor for changing those ecological relationships. Pretty much anything we do management-wise is going to influence that soil ecology. In ecological farming, we're really managing an ecology. By learning what your soil is—what your biology is—you can control most diseases through management—unless you're doing something that is very detrimental or has long-term effects on biology.

JOHN: That's a very impressive statement. We can manage disease and pathogenicity based on how we manage our soils, from a cultural perspective. That's a very, very important perspective that I think we don't commonly hear in agriculture.

You mentioned a number of different management tools: crop rotations, using cover crops, tillage, the impact of moisture, etc. Earlier, you spoke of the differences between soil-borne pathogens and soil-inhabiting pathogens. It's fairly well understood that we can use crop rotations to manage soil-inhabiting pathogens. Are you suggesting that it's also possible to use these tools to manage and suppress soil-borne pathogens?

DON: Very definitely. Most of our soil-borne pathogens have very limited genetic resistance. We rely on those management techniques to control them. Sometimes we don't recognize it as much as we need to, but soil-borne pathogens have a much more limited relationship as far as population dynamics. We may measure population of spore load and other things for organisms like *Fusarium*, but the organism is there in a high enough population that regardless of what we do—if we didn't have the other organisms associated with it—it would take our crop.

Soil colonizers colonize only as long as they have a nutrient base to function with. So we can either extend the time between susceptible crops—which we typically do with most of our potato pathogens, for instance—we see them building up in two or three crops and we want to break that population down. Same thing with anthracnose on corn. It's a soil colonizer. *Cephalosporin* on wheat. All of those organisms survive in the residue. Many of them even produce an antibiotic, so they slow down resident degradation to extend their lifetime in the soil so that other organisms aren't able to colonize that food base.

This is quite different from *Rhizoctonia* or *Fusarium*—many of the *Basidiomycete*-type pathogens are very excellent soil inhabitants. They don't require the base of nutrients that many of our colonizers do.

JOHN: You spoke briefly about the use of crop rotations and that 85 percent of the effect, in terms of disease suppression, happens from the prior crop or the prior cover crop. What are some particularly useful crops or cover crops that have a very strong disease-suppressive effect?

DON: That's going to depend on your disease and your overall soil biology. For instance, if you're dealing with take-all, *Gaeumannomyces graminis*—the root and crown rot of cereal crops—you'll find that brassica species have a suppressive effect. Perhaps the best cover crop overall is oats—another cereal crop. When we bred crown-rust resistance into oats, this also gave us an oat crop that provided disease control—take-all control—for our wheat and barley.

The reason is that crown-rust-resistant oats also produce a glycolcyanide root exudate that suppresses the manganese-oxidizing organisms. If you suppress the manganese-oxidizing organisms, you also suppress the manganese oxidation by the pathogen that is required for virulence. You've increased the manganese availability for the plant—for its own resistance.

The shikimate pathway is a pathway that gives tolerance or resistance to take-all because that's where the ligno-tubers are formed. Lignification and callousing—all of those materials are produced through the shikimate pathway. And manganese is a very critical component in that pathway—at six or seven different steps in the pathway. If you inhibit the availability of manganese—if you have a good, strong mineral chelator that ties up manganese—you're going to increase take-all because you reduce the functional availability of manganese for the plant in its own defenses.

A plant like rye is very efficient in the uptake of manganese and other micronutrients. Rye takes care of itself with its resistance to take-all pathogen, but it doesn't do anything for a subsequent crop. It does very well, with very little disease pressure, because it's very efficient in taking up manganese and other micronutrients. If you have triticale, which is a wheat-rye cross, if it doesn't contain that section of the rye chromosome that is responsible for micronutrient uptake, then the triticale will be as susceptible to take-all as a wheat crop.

You don't get a crop-rotation benefit out of rye like you do from oats. The root exudate of oats has a very strong antimicrobial compound against the manganese-oxidizing organisms that make manganese less available. You'll see that effect—that change in the soil biology—carry on for two or three wheat crops after an oat crop. Subsequent crops will have very little take-all. Oats probably have the most dynamic effect in this regard.

Brassica species—canola or mustard—also produce quinolones and some other materials that have a similar ability to reduce take-all as the glycoprotein in oats. Following canola, you'll see an increase in some of the other diseases. It's not just influencing one particular disease. If you're using Roundup ready canola, genetically engineered canola—where you're adding a very strong mineral chelator, because that's how glyphosate works, by tying up those minerals in the physiology of the plant—if you're growing Roundup ready canola and applying glyphosate, it's going to move out of the root exudates and change that soil biology. Then you may see a reduction in take-all, but you see a very dramatic increase in *Fusarium* root rot, as well as *Fusarium* head scab and the toxins of that particular plant pathogen. You're changing the dynamics of the system with the particular management tools that you use.

JOHN: When you speak of the disease-suppressive effects of oats and brassicas, what are the key characteristics that they share that perhaps are also shared by other disease-suppressive crops? What are the various crop types, or what are the characteristics that they can use to help develop a disease-suppressive soil?

DON: Many of those are worked out from empirical information, but it all boils down to nutrition one way or another. It may be indirect—through that antibiotic effect, through root exudates, on a particular group of organisms—or it may be a direct effect—increasing the availability of a specific nutrient. I mentioned the take-all situation with manganese. You'll see the same relationship of manganese with potato scab and with verticillium wilt. There's a little different involvement in the physiology with verticillium wilt—you have a vascular pathogen versus a cortical pathogen. But the availability of that nutrient or its nonavailability determines the disease severity.

People are concerned about the loss of the Cavendish banana because of the resurgence in Panama disease—*Fusarium cubense*. And yet that pathogen has always been there. We've changed the management and the soil ecology, and that provided an opportunity for that pathogen to become more virulent and to manifest itself. We've changed it from a saprophytic organism, essentially, into a very virulent pathogen, and we haven't recognized the stages that are involved in that shift. In this situation with the Cavendish banana, you have a tie-up with zinc and, again, the involvement of a very strong mineral chelator.

JOHN: You spoke of the effects of tillage on developing different microbial profiles, and I wanted to clarify that. Are you suggesting that all tillage applications provide a rebalancing effect on both reducing organisms as well as oxidizing organisms?

DON: I am. And it depends. It doesn't have to be every year, but there are some changes that take place with long-term no-till that reduce the efficiency of the ecology—for nitrogen especially. You'll have tenfold greater nitrogen reduction—denitrification—with no-till than you have with conventional tillage, even though in conventional tillage, we have a more oxidative environment. You'd think it'd be the opposite, but it isn't because denitrification is an anaerobic process.

If there's any way for that nitrogen to be converted to nitrate nitrogen in a no-till situation, it's going to be very rapidly converted to elemental nitrogen and moved back into the air. Nitrogen oxides and other materials will evolve much more extensively in no-till than in a conventional system. It's a balancing situation, and there are several studies that have shown that in long-term conservation, tilling and some of those practices—where you haven't had any soil disruption for eight or ten years—just a single shanking through the soil will increase soybean yields 50 percent.

You don't have to do that every year. You don't have to beat your soil to death to get those effects. It's just a matter of getting enough shift in that soil biology, and enough activity, and we get the desired result from a nutrient standpoint.

JOHN: What are the impacts of nitrogen and nitrogen applications on developing disease-suppressive soils?

DON: Most of your soil organisms are hungry for two things. One of them is nitrogen. The other is carbon. When you change either of those nutrients, you see tremendous stimulation of a lot of organisms in the soil, depending on what your source of nitrogen is.

One of their primary pathogens on a lot of vegetable crops in Florida is *Fusarium oxysporum*. It's a vascular *Fusarium*. Growers can get pretty much complete control by using nitrate nitrogen and calcium. If they can stimulate nitrification, or if they apply nitrate nitrogen, potassium nitrate, or calcium nitrate—and also use some liming to get their pH up—they have fairly effective control of *Fusarium* wilt diseases.

With tobacco, where we use fumigation to control some of the soil-borne diseases, you should have at least 30 percent of your nitrogen as nitrate nitrogen. If you don't—if the plant is taking up all ammonium nitrogen—you can get into a carbon deficit because the plant detoxifies the nitrate or the ammonium nitrogen by combining with photosynthate from photosynthesis. That provides the carbon base for those amino acids. Then the nitrogen is translocated as the amino acid. If you don't have enough nitrate nitrogen present to buffer against an ammonium source—if you're going to get fumigation, because our soil fumigants tend to knock out nitrifying organisms—or if you don't get nitrification there, it stabilizes in the ammonium form. It can be a drain on the carbon and the energy availability until that ammonium is detoxified and utilized by the plant.

JOHN: I'm thinking about your description of how many soil-borne pathogens—soil-borne fungi such as *Fusarium* and *Verticillium*, for example—are dependent on oxidizing manganese and limiting manganese absorption by the plant. If that were the case, what would be the impact of adding ammonium to such an ecosystem—where you have a reduced form of nitrogen? What would the impact of the ammonium be on these soil-borne pathogens?

DON: On that group of pathogens, you will see a tremendous reduction in disease—with the ammonium nutrition. I wrote a chapter in one of the annual reviews on the impact of the form of nitrogen. You'll see a tremendous benefit in different organisms by the form of nitrogen that the plant is predominantly supplied with. If you modify the environment so that those soil microorganisms make those conversions, you'll see that the form of nitrogen will be available for the plant.

It's important that you have both the management tools as well as the form. Most of our soil's nitrification takes place very rapidly, so you need to do something to inhibit nitrification. We can do it biologically—we can modify the speed of that reaction so that we can increase the amount of nitrogen in nitrate or ammonium. When it's taking ammonium nitrogen, you have a reducing environment. That reducing environment is favoring manganese-reducing organisms so that there will almost always be an increase in manganese availability for the plant—when you predominantly use an ammonium form of nitrogen. The more rapid, oxidative form is the nitrate source of nitrogen.

JOHN: What is the impact of carbon-to-nitrogen ratios on both disease-suppressive soils and also on yield?

DON: The carbon-to-nitrogen ratio depends on the carbon source.

That got me in trouble with my first publication in plant pathology. I challenged the carbon-nitrogen ratio hypothesis. People were saying, "If you have a 12:1 versus a 40:1 ratio, you'll always have a disease relationship." I demonstrated that it's not the carbon-to-nitrogen ratio; it's the form of nitrogen that is involved in that ratio.

You can take different crop residues, or a different cropping sequence, and it's the effect of that sequence on the form of nitrogen that determines what the disease reaction is. Of course, the effect of that form of nitrogen quite often is an effect on manganese or zinc or copper or other nutrients, along with the form of nitrogen.

Carbon-nitrogen ratios work if you're working with the same nutrient source or crop residue and then varying the nitrogen ratios by either harvesting plants when they're greener or harvesting plants drier—when you have wider carbon-nitrogen ratios. But the carbon-nitrogen ratio per se isn't the factor that's involved there. It's the effect of that ratio on the form of nitrogen and the other minerals that are involved—such as manganese or zinc or iron or copper—that are critical for particular physiological processes.

JOHN: Don, one of the things I believe is quite important that we haven't spoken about is the general impact of photosynthesis and the quality of photosynthesis—how photosynthesis can vary in crops and cover crops and how that influences the volume of root exudates. How can a grower increase the quantity of photosynthesis and increase the quantity of root exudates in the soil profile?

DON: You're not going to have any photosynthesis if you don't have manganese. Manganese is critical for splitting water; it provides the hydrogen that can then combine with carbon dioxide. You're not going to have any photosynthesis without magnesium, which is part of the chlorophyll molecule. You're not going to have a very efficient photosynthesis without iron and sulfur and all the other minerals because your physiology is all tied together.

If you want to improve the efficiency of photosynthesis, the first place to look is mineral availability—having that system work. If you don't have a backlog of sugar as fructose or glucose, you want that sugar to be stored as sucrose. That changes the osmotic relationship; it changes the overall physiology of the plant. You're also not going to have any sucrose if you don't have manganese because manganese is responsible for your sucrose-phosphate synthase enzyme as a cofactor.

It's a system that works together. If you don't have sulfur, you won't have enzymes because most of your proteins are initiated with either cysteine or methionine—your sulfur amino acid. C4 plants have a more efficient photosynthetic pathway. They have PEP carboxylase, as well as rubisco enzymes, after the carbon dioxide from the air binds with the hydrogen that is split off of the water by manganese. You have C4 plants and C3 plants and the physiology that's involved, but all of them require the mineral nutrients. Any one of those deficiencies influences the overall efficiency of the whole process.

JOHN: In the research work that we have done in the field, looking at various crops, I have suggested that what we have come to accept as common or normal—depending on how you define "normal"—our crops are photosynthesizing at only a fraction of their inherent capacity. Perhaps as little as 25 percent in a few cases or even less than that. What are your thoughts? Do you see the possibility for plants to increase photosynthesis volume much more than what we normally anticipate?

DON: We're only affecting the tip of the iceberg of the potential of that plant. Probably the most limiting factor is carbon dioxide. If you increase carbon dioxide just a small amount, it'll have an exponential increase in photosynthesis. This then gives us the oxygen that we need.

The system's tied together. If we limit carbon dioxide, we limit the amount of oxygen that is available for our growth. We can vary the structure—the morphology of the plant—from a leaf standpoint to capture the sunlight more efficiently. It doesn't do much good to do that if we don't have the carbon dioxide available. In a corn plant, for instance, with 250-bushel yield potential, 20 percent of that carbon dioxide is coming from the soil. But if we have a sterile soil, we don't have respiration going on in the soil. If we can't utilize that organic matter that we've applied to the soil, from which we get respiration and carbon dioxide, we compromise photosynthesis because there's not enough carbon dioxide in the soil to produce high yield potential—and the quality of the crop that we want—without that additional carbon dioxide coming from the soil and the turnover of organic matter.

Organic matter breaks the soil. We want to maintain it for a number of reasons, but if it's just sitting there as straw or stubble—not decomposing, as we see in some of these corn fields, where all of the residues are just sitting there tearing up tractor tires—it's not doing very much good from a production standpoint. We want it to break down. We want humates and fulvic acids and those materials that give us overall ecology in the soil—that give us crop production.

JOHN: One thing I often say—which really captures people's attention—is that one of the important reasons to have soil organic matter is so that we can lose it.

DON: That's correct. We're so limited on carbon dioxide, and our C4 plants are so efficient, and yet we're not tapping into the efficiency of the plants because we limit our carbon dioxide. If it's not working for us, it's really working against us from the overall system standpoint.

JOHN: When we look at the horticultural space—indoor greenhouse production—when we increase CO_2 levels to 1,100 parts per million, that increase—with the exact same lighting and exact same environmental conditions—produces about double the plant biomass just because of increased photosynthesis.

DON: That's correct. Our limiting factor in many of these situations is carbon dioxide. It's our most limiting mineral.

JOHN: What is the potential for plants to increase the volume of photosynthesis?

DON: The potential is 100 percent. I mean, fivefold, tenfold—depending on where we are now and what plant we're talking about. There are a number of different ways to do that.

I mentioned plant morphology, but you can also do it by increasing light intensity. You got a lot of leaves on a garden plant—a mature corn plant at tasseling time—that aren't getting very much sunlight. They're not photosynthesizing at their potential because of that lack of sunlight. That's where morphology comes in.

Plant spacing—we can increase the number of ears on a corn plant. There is tremendous work being done at Purdue, for instance, in corn breeding/genetics. They're producing five and six ears on a corn plant. If you increase the efficiency of that plant, you don't need as high a population. You can minimize that crowding and that shading effect.

We can reduce the allelopathic effect of our plants. To increase the population in corn, it's very critical to have even germination. If you delay germination six to eight days, you automatically reduce the size of your ear by the allelopathic effect—that autointoxication effect of the root exudates on an adjacent plant—by as much as 80 percent.

JOHN: Wow!

DON: When the John Deere MaxEmerge planter came out, I remember how it was almost an instant success because it increased the soil-seed interface—the contact that gave you that uniform emergence to minimize that allelopathic autointoxication suppression that you otherwise had from those higher populations. We got an increased corn population and maintained full yield potential without the allelopathic chemicals reducing the overall production potential of that particular plant.

There are a number of things we could do if we needed to. Right now, we're concerned about a surplus. The biggest thing is that necessity is still the mother of invention. But there's plenty of potential there.

In one of our brainstorming sessions at Purdue, we asked Charles Tsai what the biochemical genetic potential of a corn plant was. A couple of weeks later, when we were all together, he said, "I've got your answer." We were shooting for 350 or 400 bushels in our research. We knew that a lot of our better farmers—back in the late seventies—had the potential for 550 or even 600 bushels. We were trying to design some systems for them to achieve that on a field basis.

Charles sat down, and we said, "Well, is 600 bushels a realistic figure?" And he said "You're all pessimists. The biochemical genetics of that plant are about 1,100 bushels." That's what we could do if we managed the environment and the plant in a proper manner and provided the expression.

We probably won't come close to achieving that until the necessity is there. The limiting factor is the innovation of man. As long as we're doing okay—as long as we don't have that burr under our saddle to look at both the genetics and the environment—we won't have the need to maximize the expression of that genetic material.

JOHN: What yields did you end up achieving on your yield trials?

DON: We could get 400 bushels. We had farmers that were getting 350. Of course, the average was still 75 bushels. They were doing three or four or five times what the average was on a major piece of land. They were able to do it because they recognized that they were managing an ecology. They started out with the soil. They had a beautiful soil.

I remember visiting Herman Warsaw's farm. You could take a steel probe, and you didn't have to lean on it. You just pushed it in the soil three feet. He had that system working very well. In some of his neighbors' fields you'd get the probe down three or four inches and you'd have to really put some pressure on it. To get it down a foot and a half, you'd be driving it in.

I can't tell you how many of the old ping tubes are still sitting out there in the field. Those were tubes that you could drive into the soil like soil samples—at three and four feet. We had to get it into our soils using a sledgehammer. Then the problem was pulling it out. You would tear the metal off of the tube with a jack, and finally, you'd just crimp it over so that it wouldn't tear up the tire, and then you left it. A lot of them are still sitting out there in those fields because we didn't have that concept—the soil wasn't a major part of the management program. In general, we would look at the nutrients and forget that all parts of the ecology needed to be managed—to have better percolation, to have biology, to have air exchange—and most things have to take place if you want to capitalize on the genetic potential of the plant.

JOHN: What were the plant populations that growers were using to achieve 350-plus bushels per acre?

DON: You had the old Pioneer 3532 seed—a hybrid that picked up 95 percent of its nitrogen by tasseling time and then merely recycled it. On sandy soils, it was a great hybrid because you couldn't maintain your nitrogen availability in those sands. It would take it up and store it, but it had a yield potential of about 125 bushels at 24,000 plants, which was our standard at the time. But you could increase the population of that particular variety because it didn't have the allelopathic effects from the root exudate with high population. It had a fixed ear. As you increased the population, you still maintained that same ear length. It wasn't bigger like your higher-yield hybrid, but it was very stable, and it tolerated high populations. They got the yield up by increasing population.

With our other varieties—our high-yielding varieties that were hybrids—there you had a flex hair that related to the environment more dynamically. The higher the yield potential, the more nitrogen you want as ammonium. With the 3532, a fifty-fifty ratio of ammonium to nitrate was optimum. If you get into the higher yield and the 250- to 300-bushel yields, you want 75 to 80 percent of your nitrogen as ammonium and only 10 to 20 percent as the nitrate source of nitrogen because you want as much photosynthesis as possible to go into the kernel. When the plant utilizes nitrate nitrogen—and most plants can utilize either form equally well—it takes 15 to 20 percent of your photosynthate to reduce nitrate nitrogen back to the amine form so that the plant can utilize it.

The higher your yield potential, the more ammonium nitrogen you want—to provide those amino acids for that growth and your enzymes and everything—and less nitrate nitrogen. We always found that there was a benefit to some nitrate nitrogen because it serves as a buffer—both to limit the drain on carbohydrates—if you have a high uptake of ammonium nitrogen, nitrate will tend to balance that—but also as a stable form of nitrogen. If you run short, then you can use some of that photosynthate.

If you have molybdenum and your other nutrients available, you have part of the function of your nitrate and nitrite reductase enzymes. Again, you have a different physiological pathway. If you're saying, "Well, I'm going to get most of mine from an ammoniacal source," you may forget that you also have to have molybdenum for some of those other enzymes that aren't quite as dynamic or quite as involved as they are if you're relying more on the nitrate source.

Those are some of the things you could do to enhance that overall photosynthetic efficiency—the form of nitrogen is going to influence your soil biology and your buffering capacity in those areas.

JOHN: Don, in 1979 you were producing 350-plus bushels of corn per acre in a biological soil ecosystem. Today, growers are struggling to produce 250 bushels of corn. We don't even have a conversation about growing 350 bushels of corn on a commercial field scale. There are a few notable exceptions, but not on a large-scale production system. What happened with that knowledge? Where did it go? Why was it not adopted on a much broader scale?

DON: We started saying we had too much production. We needed to focus on different things. At our land-grant universities, a lot of that research and the long-term commitments that breeding programs require for the expression of that genetic potential was closed out. Materials were just given to the private companies to develop their experiment stations.

The universities were happy to not have that long-term commitment. They could then respond to the political pressures, and their programs started being limited to three to five years—for the competitive grant programs on a federal scale. And most of our breeding programs were funded through the Hatch Program and the Smith-Lever Program, which would give the states a constant amount of money on a formula basis for those long-term agricultural developments, which are the reason why we have success in our agricultural programs. They were built on those long-term, continuous programs that were pretty much abandoned as we started looking at the bells and whistles in science rather than at the end product.

Again, we were producing more than we knew what to do with. I don't know what we'd do with all the corn that we currently produce if we weren't producing so much ethanol. I mean, that's the way to use your crop: find a new market for it. Certainly, population growth is a long way from requiring our current production. We could produce enough food for about fifteen billion people with about 30 percent less land—if we wanted to really do that, if we really needed to do that—with the technology that we had in 1964.

We were shooting for 400 bushels in 1979 and 1980, and now we're struggling with 250 bushels. But sometimes you have to reinvent the wheel. That part of the system was not considered important, and the resources were fractured. In a breeding program, you don't just turn it on and off with each little whim or political idea that comes along. It's a long-term program. When we turned all of that material over to the private companies, their interest was the bottom line. There's a tremendous amount of material that could be manipulated. As far as that long-term commitment, there hasn't been any of that.

Genetic engineering certainly has not improved the long-term effects; you get the idea that we can do it all in a laboratory just by switching this system on or inhibiting this particular system. We forget that it's still a system—an ecology that has to be managed—if any of it's going to be of value to us. It's a thought process that's involved, as well as the necessity. But also, the desire—the innovation—drops out when you forget that you're a part of a very dynamic, beautiful system that was all put together—when you start focusing on only one thing. Silver bullets may take care of a varmint, but they don't provide stability in the system.

JOHN: Don, what is something you believe to be true about modern agriculture that many other people don't believe to be true?

DON: I think there are a lot of people who feel the same way I do. They may not be quite as outspoken as I've been.

When you focus on just one item and forget that all of the systems work together, you automatically compromise your potential. Sometimes we have to realize that you have to make a few mistakes in order to do that. We're making plenty of them now, I think. Maybe that's part of the gray hair that I carry around with me.

Also, I see a lot of a light at the end of the tunnel. The tremendous damage that we're doing really hurts. To think of your grandchildren having a shorter life expectancy than I do, as the CDC has stated—it doesn't have to be that way. Sometimes you have to smash your thumb to realize that maybe you needed glasses to find the nail.

You've been successful in communicating the importance of managing the system rather than just looking at the bottom line. Putting it in perspective is what we need to do. I see some young people that are able to express that a lot better than I am.

JOHN: Don, what do you see as the biggest opportunity in agriculture today?

DON: Reinventing the wheel. The greatest need, anyway, is getting nutrient density back and eliminating the pesticides—especially glyphosate. Jack van Brook's research in Washington State says that on a 0 to 100 scale, from a health standpoint, glyphosate is about 82. All other pesticides together, other than the neonicotinoids—which would be at about 14—are less than 1.

JOHN: Wow, that's incredible! I was not aware of that.

DON: We focus on a lot of the chemicals that are less than 1. Not that we shouldn't, but we totally ignore the one primary material that is doing the greatest harm from an environmental standpoint. And that's because glyphosate is not just a pesticide per se. It's an antibiotic. It's a strong mineral chelator. It has the ability to essentially disrupt everything that we value in our ecology and for our own subsistence.

It needs to be removed from our soil and from our food because of that wide scope of damage that it's doing. I'm not opposed to tools in agriculture and making things easier. I remember when 2,4-D was available to us after World War II. What a tremendous tool we thought it was. But it also required some common sense and some recognition. We don't have silver bullets.

We have an ecology that we need to manage. If we manage it effectively, it will serve us for eternity. If we don't, then we have to go through correction phases that can be very painful for us—self-fulfilling prophecies of food shortages. You look at the recent elections. You see the change in the German government that took place. That was a surprising change for many people. It took place because of agriculture. You can go through history and show how important agriculture is. It's the basic infrastructure of any society. When we foul it up, there are consequences, and it requires a correction.

I think we're seeing the pendulum swinging now. We're recognizing the need for nutrient density and the need to eliminate a lot of the exposure that we have to glyphosate and to some of the other chemicals that have been accumulating in our environment and our food supply. We'll see those changes taking place. Not as fast as I'd like, but they're coming. I don't move very fast either, so I can't necessarily complain.

It's critical that we recognize the stewardship we have when it comes to the soil—and our own survivability—rather than just accepting the changes and saying, "Well, there's nothing we can do about it now." There's a tremendous amount we can do. It's a matter of understanding that we're managing an ecology.

JOHN: Don, what are some books or some resources that you would recommend to growers who want to learn more?

DON: You have the American Phytopathological Society, under the direction of Lawrence Datnoff. They published the book *Mineral Nutrition and Plant Disease*. If you're interested in those relationships, I think that's a good book. I'm one of its editors. And you have Marschner's book—*Mineral Nutrition of Higher Plants*—which is probably the Bible on plant nutrition of plants. It's a little detailed—a little deeper than a lot of people want to go. You have a lot of the older books—the Albrecht System.

There are a number of resources. You have your programs that you're working on. You're able to communicate with growers. There are eight or ten different systems now looking at nutrient density: How do we measure it? How do we define it? They're at least getting a lot of that background information that's necessary for us to really make the application.

We recognize that in scientific accomplishment there are some unintended consequences. If we don't recognize that science is merely a tool—if we rely on it solely without recognizing that there are unintended consequences—then it can come back and bite us pretty hard.

I think that's what we're seeing in our general adoption of genetic engineering—without recognizing the unintended consequences that are built into that system, either from a mutation standpoint or in the direct damage that is done. There's a reckoning process that we all have to go through. We're going through it now if you're looking at the health consequences. All you have to do is go to a cancer tumor center or a child development center or the maternity ward and see the birth defects, which are becoming so prevalent that you can't ignore them.

We're probably in that reckoning stage. It becomes important that we start looking back in the box—rather than just outside the box—for the solutions that we need for society.

JOHN: Don, what is a question that you wish I would have asked?

DON: We didn't get too much into the GMO-glyphosate conversation. That takes us in a little different direction—a very critical one when it comes to nutrient density because of its impact on soil ecology as a whole—our entire ecology. Glyphosate is such a strong mineral chelator, and it has an antibiotic effect. The antibiotic effect may cause greater damage than the direct chelation. It's a broad-spectrum antibiotic because it's a broad-spectrum chelator.

I'm really happy to see all of the emphasis coming back to soil ecology—the role of different crops and the rotation and the overall cropping system and the use of cover crops. I'm really excited to see that. That's something that sixty years ago, when I started in plant morphology—looking at the effects of crop residues and crop sequencing on soil-borne disease—I thought there had to be a better way than saying, "Well, you can only grow a primary crop every eight years because the disease pressure is going to build up on you to the point that it's not going to be economical."

We could demonstrate that very easily back in the 1950s and 1960s. Then in the remediation aspect—maintaining the balance in that soil environment—the fact that 85 percent of that effect comes from the immediately preceding crop means that you can essentially do the same thing we were doing in a five- or an eight-year rotation at the same time that you're growing an economic crop. And even doing better—primarily if you look at the improvement in supporting that growing crop from a nutritional standpoint—nitrogen or other minerals through the form of nitrogen that we can influence with that particular crop's root exudates.

I think those things are really exciting. Other people don't seem to get quite as excited as I do, but it takes me back to those questions that I had: How do we do this better? If these are things that we need to do from a biological standpoint, isn't there a way to compress some of that process and capitalize on the potential dynamics of it? We see the end product on a consistent basis rather than just the ups and downs that we would have otherwise.

The Value *of* Regenerative Agriculture Management

JERRY HATFIELD

Regenerative agriculture focuses on the fact that we have to improve our soil resources. It shows how we can do that, how changing the soil benefits that plant, and how, in terms of productivity, it benefits the ecosystem—in terms of improving not only production but also environmental quality.

In regenerative agriculture, we're admitting that there are a lot of interactions, and we aren't afraid to say that changing something in the soil is going to have some impact on the plant, and it's going to have a different reaction depending on the weather.

I always tell producers that biology wants four things. It wants food, water, air, and shelter. These are the basic necessities of life that you and I want. We should start thinking about biology from that perspective.

We use a concept called "genetics by environment by management." I always explain to producers that management is what they oversee and the decisions they make. Environment is what they're trying to overcome—the soil environment, the weather, etc. And genetics is what they're trying to optimize.

REGENERATIVE AGRICULTURE is becoming much more mainstream than it was even three to four years ago. There is a rapid development of new ideas, new technologies, and new information, as well as rediscovery of old information. For Dr. Jerry Hatfield, the rising stature of regenerative farming is a vindication of his many years of effort in the field.

Jerry received his PhD in agronomy from Iowa State University in 1975. He is currently the Laboratory Director and Supervisory Plant Physiologist of the USDA-ARS National Laboratory for Agriculture. He has authored or coauthored nearly five hundred peer-reviewed papers and has received numerous awards for his academic and research work. Jerry served as the president of the American Society of Agronomy in 2007 and was inducted into the USDA-ARS Hall of Fame in 2014.

JOHN: Jerry, you've been working on a lot of things since we spoke last year. What has been exciting for you?

JERRY: A lot of things have been exciting. I get up every day, John, with excitement in terms of these discoveries—how we have to change our thinking in soil management—how we have to change our thinking about how plants feed into this whole soil environment.

I've worked on this soil-plant-atmosphere continuum for a long time, and these last few years have been eye opening in terms of understanding the dynamics of how soil and plants and the atmosphere all come together. It's moving us forward in looking at agriculture from an entirely different perspective. It gets us to the point of regenerative agriculture really paying us dividends in terms of how we look to the future. How do we look at the future in terms of providing nutritious food for the whole world's population?

JOHN: Let's focus on that. Can regenerative agriculture provide enough food for a growing world population?

JERRY: I definitely think so. I think regenerative agriculture is the key to food production and to food security.

If we want to think about increasing our production on a unit of land, one of the key pieces is how to enhance our soil so that we can have greater productivity. But at the same time, how do we enhance our soil and production system so that we are increasing the nutritional quality of that food?

I think that becomes the key, in terms of how we look at the system from an entirely different perspective. I think regenerative agriculture is the platform that we need to think about going forward, in terms of being able to accomplish this. Not only within the US but around the world.

JOHN: Jerry, just five years ago—or even less than that—the term "regenerative agriculture" wasn't even in the mainstream lexicon. I've been very surprised by how rapidly it has gone mainstream in the last couple of years. How has that shift in lexicon shifted the thinking and adoption? There's been discussion about sustainable agriculture for quite some time, but how have actual practices in the field changed? And how can regenerative agriculture deliver something for us that sustainable agriculture wasn't able to?

JERRY: That's a really good question. The whole aspect of sustainable agriculture was that we were going to build agriculture for sustaining production and sustaining economics. The National Academy of Sciences put out their big report on sustainable agriculture.[2] But it really didn't address how to get there.

Regenerative agriculture focuses on the fact that we have to improve our soil resources. It shows how we can do that, how changing the soil benefits that plant, and how, in terms of productivity, it benefits the ecosystem—in terms of improving not only production but also environmental quality.

I think it's really about the mindset of how these pieces come together. It really could be as simple as the fact that sometimes it takes society a long time to reach a certain level of maturity.

It's really caught on because, I think, people can begin to see the pathway forward. People stood at the sidelines in sustainable agriculture and said, "Yes, that's a nice concept, but it doesn't have a road map."

Regenerative agriculture is supplying the road map.

JOHN: I've made that comment in the past: Why would we desire to sustain our current levels of productivity and the externalized costs that we have?

When you think about the differences between sustainability and regeneration in the local field—how do those differences really play out, in terms of management and results?

JERRY: That's a good question, John. You're right. I always had an objection to calling it "soil conservation" because in soil conservation, you just conserve

---

2 *Toward Sustainable Agricultural Systems in the 21st Century*, National Academies Press, https://www.nap.edu/catalog/12832/toward-sustainable-agricultural-systems-in-the-21st-century.

what you have. Even back thirty years ago—when I first came here to the lab—I said we needed to be talking about soil enhancement. How do we improve our soils? How do we get more carbon into them? How do we get more water into them? How do we get better nutrient cycling?

The point of regenerative agriculture is that we're going to build up a soil base. We realized that soils are different within fields—that agricultural systems are going to have different linkages with the soils and the climate in which they're growing.

We began to put it together in a platform that said, "Here's a unique combination of soils. Here's a unique combination of what plants we grow there and the climate that's there." We've been doing a lot of work lately on agroclimatic indices—what plant will fit where, with the most sustained production but the least variable production over time.

It always comes back to the soil. If I want to reduce my variation and make the best expression of the climate, I've got to have soil that's supplying water, nutrients, and support to that plant, with everything linked together.

Science is really reticent to dive into a lot of the interactions that we know go on within an individual field. We've got plants, we've got insects, we've got diseases, we've got soils, we've got weather. All of those are varying in time and space and are interacting together in ways we can't comprehend at this time. We try to quantify them, and we get even more mystified.

In regenerative agriculture, we're admitting that there are a lot of interactions, and we aren't afraid to say that changing something in the soil is going to have some impact on the plant, and it's going to have a different reaction depending on the weather.

JOHN: I think that's a key challenge that we have with the scientific method and single-factor analysis—that agriculture doesn't lend itself to single-factor analysis very readily, or even multifactor analysis. There are almost an unlimited number of factors.

JERRY: That's what makes it exciting—because there are an unlimited number of factors.

An example of this is the fact that rising CO_2 levels are extremely beneficial to plants. Farmers love higher CO_2. But at the same time, as we increase temperatures—or make temperatures more variable or make precipitation more variable—we negate that CO_2 effect. These interactions between water availability, temperature dynamics, and rising CO_2 are very critical for us to look at.

I think we're now mature enough to say that we can't just look at one parameter alone. We have to look at all of these things as a system. Producers have been doing this since they started cultivating. There are a lot of times when growers have been very frustrated with scientists because we scientists don't see the same big system the growers manage. We look at it as "What did you change in terms of nitrogen? What did you change in terms of phosphorus? What was your row spacing or plant population?" It's all of those things. We scientists are going to have a look at this and develop a strategy on how we go forward.

JOHN: When I think about the differences between sustainable agriculture and regenerative agriculture, the approach we've taken in our work is that regenerative agriculture is all about helping plants get to peak photosynthesis, produce an abundance of carbohydrates, and move those carbohydrates into the soil, where you have a very functional carbon cycle working. You're constantly accelerating plant health and constantly accelerating soil health.

What I've realized is that when we think about the entire soil-plant system—as an ecosystem—you have photosynthesis, which is the way you bring new energy into the ecosystem, and you have soil biology, which processes that energy into soil. And that total energy flow—voltage, if you will—is reflected in the carbon cycle, the carbon exchange.

How does the carbon cycle shift and change when growers begin managing soils and crops differently—with regenerative management rather than with present mainstream management?

JERRY: We do a lot of work in corn-soybean systems. Over the past seventeen years, looking at the exchanges of carbon between the plant, the atmosphere, and the soil, we've shown that our typical corn-soybean system is losing a thousand pounds of carbon per acre per year. This is with maybe a deep rip in the fall and field cultivation in spring, and the only thing taken off the land is the grain of corn and soybeans.

Think about the average life of a producer. Farming forty years, they've lost forty thousand pounds of carbon—twenty tons. It's a slow loss, but that slow loss is impacting the aggregate stability of their soils. The farmer realizes he's losing productivity, that it's different than it was before, but he doesn't come to the realization that it's the cumulative effect of what he's been doing over his farming career.

That's a dynamic that we need to consider when we talk about why fields become variable overtime. But on the other hand, we can change that system quickly. I think this is the framework that we need to be talking about. When we go to a system where we add cover crops, and we reduce the tillage intensity by going to strip-till or no-till, we find that within one year, we can change that negative carbon balance into a positive carbon balance. Then we can put more dividends from that plant back into the soil biology. We go from a negative to a positive carbon balance.

JOHN: In a single year? That is amazing.

JERRY: Yes, and even more amazing is that over a two-year period, we doubled the microbial biomass in the upper twelve inches of that profile. These are not test plots. These are 160-acre fields, and they're sampled at 150-foot grids. There are a large number of samples coming out of those 160 acres. We were able to improve the biology very, very quickly.

We've already begun to change the upper surface of that soil. The cover crop is giving us a longer period of time in which to take carbon dioxide out of the atmosphere—converting it into carbohydrates, putting it back into the soil, and feeding that soil biology.

I always tell producers that biology wants four things. It wants food, water, air, and shelter. These are the basic necessities of life that you and I want. We should start thinking about biology from that perspective. They want a food source, just like you and I like to eat every day. In a lot of our systems, we were only growing a crop during the summer. There was a long period of time on either end of the growing season when we weren't feeding the soil biology. It had to exist on what was there, and that's pretty much a starvation diet. If we didn't eat three months out of a year, we'd probably be fairly thin.

Cover crops utilize a lot of solar radiation. If we don't have a crop growing, that sunlight is just going into the surface of the earth and isn't doing anything for us.

JOHN: It's actually having a negative effect.

JERRY: Yes, it actually has a negative impact.

The other piece of this that we're very excited about is a small grow-chamber experiment in our laboratory in which we've been growing a monoculture of oats and then a cover crop cocktail with an oat, a legume, and a brassica species.

We see that within one growing season, the cover crop has already changed the aggregates in that upper soil surface. We've already seen it change the carbon cycle. In the cocktail, we see a much higher CO_2 level, which is indicative of biological activity. We see a better oxygen content in the air—we've opened up the pores, so we get better oxygen exchange between the soil and the atmosphere. We see that we've changed the aggregates. We're putting more carbon in there. After the first year, they are not very stable aggregates. They are starting to form.

The more fascinating thing is that we terminated those cover crop experiments by cutting them off and laying the residue back down on the ground, mimicking a roller-crimper system. We planted soybeans behind it, and we found that we had enhanced productivity when we used the cover crop, but we had an even more enhanced productivity when we used the cocktail—more and bigger beans.

Then we went back after the soybeans and planted the cover crop cocktail again. We grew that for a season, terminated it, and planted grain sorghum. We're seeing that the plants that are in the cocktail are much greener. We're running SPAD meters to look at leaf chlorophyll content. The lower leaves on the cover crop and the cover crop cocktails have remained green, while the leaves without the cover crop are already starting to dry up.

It isn't only the carbon cycle—I think cover crops are influencing the nutrient cycle. They're supplying nutrients continuously throughout the whole growing season.

JOHN: Jerry, when you spoke about growing in grow chambers, you mentioned that you saw an increase in oxygen content, as well as CO_2. This is something worth elaborating on because I've observed the same thing in the field when we have plants that are photosynthesizing well.

Many people have this idea that there is a conflict between CO_2 concentrations and oxygen concentrations. I don't see it that way at all. When we have higher CO_2—specifically when we have higher CO_2 being released from the soil—and when we have good photosynthesis, we get higher oxygen content in the air.

JERRY: That's correct. That's counter to our thought process. You only see that when you start measuring both things simultaneously. I've been looking at the literature on the number of papers that actually measure oxygen content within the soil—even the CO_2 content within the soil. It's really pretty sparse.

It's not a trade-off between oxygen and CO_2. A good biological system generates more CO_2 because the soil has more pores and more structure to allow gas exchange to occur. That keeps our oxygen content high, which then promotes more biological activity, which generates more CO_2.

JOHN: The reality is that soil and plant ecosystems are cyclical. They are self-perpetuating, and they will self-perpetuate either in a regenerative manner or in a degenerative manner. The only determining factor between those two are the decisions that the farm manager makes.

JERRY: Yes. That's why we use a concept called "genetics by environment by management." I always explain to producers that management is what they oversee and the decisions they make. Environment is what they're trying to overcome—the soil environment, the weather, etc. And genetics is what they're trying to optimize. You want to look at that genetic material and say, "What can I do to get the most out of that genetic potential in that crop? What role did my management practices play in allowing those genetics to be optimized?" I want them to think about "G × E × M": genetics by environment by management.

JOHN: Earlier, you mentioned how increased CO_2 levels in the atmosphere lead to greater climatic instability and more climatic extremes—precipitation events, etc. How do we build greater climate resilience into our ecosystems?

JERRY: One of the things that builds climate resilience is water availability. We see water as probably the limitation to biological activity, whether that's microbes or plants. Water and oxygen are our real lifeboats. The key to an agricultural ecosystem is having water available to promote biological activity.

Water-stressed plants don't photosynthesize very effectively or much at all. Even when they have plenty of sunlight and the right temperature, if they don't have water, they won't start to grow. Resilience within that system is really about managing our soils to make sure that water availability is optimal.

Going back to our discussion about CO_2 and oxygen behavior within the soil: those pores that are allowing oxygen to get down into the soil are the same pores that are moving water from an irrigation event or a rainfall.

I always tell producers that it's not how much rain they get in a rain gauge—it's how much rain they get into the soil. If we're protecting that soil surface and we've got all those pores open, we can infiltrate a great deal of water. The more we can make that aggregate stable, the less the forces of water can disrupt it. We can maintain those channels for a longer period of time—almost forever—in good soil. Once I get water into the soil, I can store it.

Farmers need to start asking whether their soil is capable of capturing rainfall events. Because when you talk about climate disruption, one of the things we're seeing is that we're shifting our rainfall patterns. We're getting more spring rainfall. We're getting more variable summer rainfall. Those storms are becoming more intense, and they're becoming less frequent.

If we want agriculture to be efficient and resilient in the long run, we've got to make sure there's an adequate water supply for our crops.

JOHN: I think there is a second aspect to what you're describing—not only making certain that the soil can store water but that it can also release it. I think this is an unappreciated quality of good soil aggregate structure because you can have some clay soil types that adsorb water so tightly to the clay colloids that they actually have no available water for plants to absorb—at 70 percent of field capacity—when you don't have aggregates.

JERRY: That's a good point. This goes back to the different soils that are out there. We want that soil not only to hold that water; we want it to release that water back to the plant.

I'm fascinated by soils that have been improving over time. They've been changing their aggregate structure. They've got organic matter around the sides of those aggregates and are very dynamic in terms of absorbing water and releasing water. When you start looking at how roots are interacting with those aggregates, you find out that root hairs—and even the mycorrhizal fungi—form a web around that aggregate and take advantage of all the little pore spaces in which water and nutrients can be stored. That becomes a real key to how we begin to look at changing soils and at our pathway toward resilience, relative to some of the weather variations that are going on out there.

JOHN: In the field-scale experiments that you're conducting growing cover crops on former corn and soybean fields—where you're getting this carbon gain—what cover crops were you using, and what have you observed there in terms of aggregate development instability?

JERRY: That was just a simple annual rye cover crop experiment. We haven't gotten into cocktails yet—that's my next adventure. We saw those changes in the aggregate structure in the upper inch of the profile. That upper inch of soil is the gateway between the atmosphere and the soil.

Improving that upper inch really does pay dividends. The more stable we can keep it—the less we disturb it so that those aggregates can continue to grow and form—the better off we'll be.

Nature loves diversity. Look at the natural ecosystem. Do you ever see a monoculture in a natural ecosystem?

JOHN: There is no such thing.

JERRY: There's no such thing. Why in agriculture do we think that monocultures are the path? Because they're easy.

But every plant gives off a different set of exudates from the root into the soil. That's what the biology is feasting on. I think we see those increases in CO_2 levels when we used that cover crop cocktail because we enriched the diet of the soil biology. We created a more diverse biological system within the soil. All of those organisms are playing against each other and complementing each other.

We understand so little about soil biology. It's really probably the biggest frontier we have in agriculture—beginning to understand what changes our soil biological system and how we maintain and increase that soil biological system.

We understand the principles. Microbes want food, water, air, and shelter, but how do those things interact to shift different bacterial species? What role do mycorrhizal fungi play, or all the other things that go on within that soil? We just can't see into the soil very well.

JOHN: And because we can't see into it very well, we struggle to understand how to value it. What is the true value of biology? I think this is a particularly challenging question in the context of agriculture because our agricultural economics are screwed up. We have an agribusiness ecosystem that has developed in which our crops have generally become commodities, operating on very low margins, combined with the challenge that we have historically externalized many of our costs. We have externalized the environmental pollution that has been caused by some of the toxins and pesticides that we've used—fertilizers, nitrates, and water, etc.

When we look at all these factors from a macro perspective, what is the value of regenerative agriculture? What's the cost of it? And what's the return?

JERRY: I've often given talks to people who've asked the value of carbon, and I tell them it's priceless. But they don't want to accept that answer.

I think we're at the point in agriculture where we need to move away from just talking about agriculture and begin to think about agroecology. How does agriculture fit into the ecological system? What's the value of its different ecosystem services? What is the impact of agriculture on water quality, on water quantity? What's the impact on the biological services that we see within the soil? What's the impact on the biological activity we see associated with that agricultural field? How do we even look at that landscape from a different perspective? I think that when we start looking at that context of agriculture, then we can really have a fruitful discussion about the value of regenerative agriculture.

In regenerative agriculture, as we improve carbon, we improve water. We also improve nutrient cycling within the soil. And not only nitrogen, but phosphorus and potassium and all the micronutrients. All of those pieces are now linked together, and those things pay me dividends.

As an example, take a silt-loam soil with 2 percent organic matter in it, and assume five-foot-tall corn in the middle of August in the Midwest. This is using water at its maximum rate. That plant has about eight days of available water before it begins to be stressed.

That's not very much. But if there's 4 percent organic matter, the plant can go for thirteen days without water. You've got five more days of available water for that plant to perform to its optimum without stress. The probability of getting rainfall during a five-day period across the Midwest is still pretty good. My point is that this is one of the ways regenerative agriculture produces value.

The other thing we see is that when we enhance nutrient cycling, we have a greener plant—a more photosynthetically efficient plant—so that more carbohydrates go into that plant. This not only improves grain production—it also continues to feed that root system so that it becomes more effective all the time.

That's where yield stability comes from. We will become less affected by the weather variation that is going on—the longer periods of time between rainfall events.

You'll always have parts of a field that are low yielding. The high-yielding parts of that field are always really good soils. The low-yielding parts are always those really poor soils that have low water availability and low nutrient availability.

JOHN: In essence, what you're describing is that we need to think about the value and the cost not just in terms of return on money but also as return on water, return on nutrients, etc.

JERRY: Yes. And let's go back to the field with low- and high-yielding parts. How many producers changed their inputs between those two different parts of the field?

JOHN: Very few.

JERRY: Very few. My contention is that our water quality problems—and I know our sediment problems—are coming from these soils that are degraded because they don't have any capacity to hold it. Water moves through them—either through leaching or run-off.

I think we can improve our water quality very simply by not putting as much on these low-yielding fields because you're not getting return on it. The other thing I've spent a lot of time now looking at is water-use sufficiency. How much crop do we get per unit of water used by that crop? You'll find out that those low-yielding parts of the field will invest a lot of water in growing that plant, but when it comes to the grain-fill stage, they're water limited; they're water stressed.

Even though we had a great harvest potential, we don't realize it because our crop is under stress and goes into early senescence. We don't capture the sunlight as well. In that part of the field there's also very low solar radiation use efficiency. We've invested radiation into that crop, but we haven't got any return on it in terms of grain yield.

Now we're not just talking about profitability—we're talking about what sort of return you're getting for every unit of nitrogen you apply. You have to couple that with water, and you have to couple that with light. You begin to see that there's a dynamic of systems in terms of utilizing their natural resources.

What are we losing in that system when we don't capture those nutrients that impact water quality? What are we doing in terms of any of the chemicals that we put on? Because we haven't utilized those in an effective way.

JOHN: Jerry, how often are nutrients—specifically nitrogen, phosphorus, and potassium—a limiting factor in that equation? When you consider water and radiation and the various components that you were describing, how often is nitrogen a limiting factor?

JERRY: Rarely. If you want to be honest about it, nitrogen is rarely a limiting factor. Although we put most of our importance on nitrogen.

JOHN: In my experience, I think I could safely say, "Never." I'd never imagine nitrogen being a limiting factor because it's the one thing that growers supply in surplus—all the time.

JERRY: As a scientist, I never use the word "never"! But you're right. Nitrogen is not our limiting parameter.

If we look at high water tables early in the season and these shifting rainfall patterns with more spring precipitation and our soil in a condition that doesn't allow drainage or the aggregate structure to allow oxygen in, then our root system may be limited to the upper two feet of the soil profile. That's as deep as those plants are going to grow. And then, when we get into the grain-fill period, there's a shift from adequate rainfall to less-than-adequate rainfall. The plant at that stage isn't going to grow any roots, and so we have a two-foot profile, and that's all the plant has to exist on. Not the whole five-foot profile that the plant could potentially use.

We see that water is a limitation to that plant. I've often made the statement that water trumps nitrogen. That's not always the most popular statement to make because people don't realize that water drives a lot of these processes and is the factor that influences productivity—more than nutrients.

JOHN: Jerry, we're talking about a complex ecosystem with lots of pieces and parts in terms of climatic inputs, soil, etc. If I'm a farmer in Iowa, where are the trigger points? What are the pieces that I can influence with the least amount of input that will have the greatest productive changes to the total ecosystem?

JERRY: One is that you have to reduce your tillage intensity because any time we stir that soil with tillage, we are releasing carbon. We're allowing those aggregates to be exposed. We see puffs of CO_2 coming out of the soil. We really disturbed the soil in that situation.

Reducing tillage can go one of two ways. You can diversify your rotations and add a crop to your corn-soybean system. You still have a monoculture in that field, but you're spreading out its rotation—maybe putting alfalfa hay into it. Or I can go to that corn-soybean system and figure out how to efficiently add a cover crop.

One of the very fascinating things is looking at how much sunlight is available to grow that cover crop in Iowa. Starting from, say, black layer on corn or total leaf drop on soybeans until the first frost—and then from the last frost until we had an emergent corn or soybean crop—we found that about a third of total solar radiation for the whole year is available within those two time periods. So we have a lot of solar radiation available to grow plant material and to put carbohydrates in the soil—in the soil biology.

For producers in Iowa, it's really quite simple in my opinion. Let's figure out how to capture more carbon from your cropping system. Let's put carbon both into something that you want to harvest and into that soil biology.

The other thing is to begin to reduce the tillage. I'm a big fan of strip-till because a lot of people don't want to no-till. Strip-till is only disturbing that small amount of soil into which you're putting your seed or your manure. Farmers are showing that there are big advantages to that in terms of improving their soil carbon.

I really believe that we can change this. We've seen advances in cover crop area across the state. We haven't seen as much of an advance in terms of reducing tillage intensity, but I think this becomes our challenge, John: How do we effectively manage a cover crop as part of the system? We also have to address the potential for an economic return of a third crop.

What's the value of putting wheat in the rotation? Farmers often say that they don't have a market for it or for a rotation with alfalfa. And what's the market for hay? I'm excited about the fact that we're talking about industrial hemp coming back and being approved. What can we do with industrial hemp in terms of paper products? It's a renewable resource from that perspective. How do we manage that?

I think that there are a lot of things that we need to be thinking about. The producers who are doing these things have shown that it works across different soils and different climates—within the state and across the Midwest. They're the bright shining examples that we can make these changes and that they are profitable but that they require some management.

It's no longer just a simple system. It requires management investment.

JOHN: There is a greater degree in management intensity, for sure. You raise a couple of questions that we need to be thinking about in terms of how to manage cover crops, etc. What is the question that growers should be asking? What is the question they should be thinking about?

JERRY: If you go back to cover crops, what's the optimum time to put that cover crop in? A lot of disruption goes on in our system; 2018 is a really good example of this. It was a very wet fall, which meant farmers had a hard time getting their cover crop established. They couldn't get into the fields if they wanted to wait until after harvest.

Contrast that to 2017. There was a really dry fall, and farmers were able to get into their fields, but there wasn't any water to stir into the surface to germinate the cover crop. All they did, in a lot of cases, was spread birdseed.

I think the question for producers is how—within their management system and with their management capabilities—to begin to identify the pathway toward improving their soils and to work to make those changes because we've shown that improving soil is going to add resilience to your system. It's going to increase your efficiency of use of your resources.

The question to the producers is really quite simple: "How do I do this?" Each answer is going to be tailored. There will be general principles, but each one is going to be tailored to an individual system.

Indicators *of* Healthy Soil

ROBERT KREMER

We're finding that some pesticides will be detrimental or toxic to just a group of microorganisms in the overall soil community. The same ones may stimulate other components of the microbial community. You end up with this imbalance of the various microorganisms that are in the soil.

There are several indicators that will suggest good soil health, but the two main ones are soil organic matter—or soil organic carbon—and microbial diversity.

Of course, a lot of it comes down to money. It comes down to the fertilizer companies trying to make money. If you tell them that some of their products are ruining some of the soil properties, that's not going to make them very happy.

We've really gotten away from what an agroecosystem should be. I think that when the family farmer or the smaller-scale farmer had livestock integrated into the system, the profitability was there. It's just that it's been hijacked by industry.

WHAT ARE THE ELEMENTS OF HEALTHY SOIL? Dr. Robert Kremer asserts that there are two main ones: soil organic carbon and microbial diversity. High amounts of active, organic carbon support the microbial community in the soil. And a multitude of different microbial species is crucial for accomplishing each of the different tasks that need to be done to support healthy plants.

Robert has a lifetime of very rich experience in the agricultural space. After earning his PhD in soil microbiology and biochemistry from Mississippi State University, he began a career with the USDA's Agricultural Research Service in Columbia, Missouri. He simultaneously taught at the University of Missouri in a variety of different areas related to plant sciences and soil microbiology.

Robert has looked at the interactions of soil microbiology and weeds, weed seeds, weed germination, and a whole range of different topics that tie into soil health and developing suppressive soils.

JOHN: Robert, you've done a lot of very interesting work. What are some of the memorable moments that have led you to where you are today and some of the things that you've worked on?

ROBERT: I think probably the first challenge was during my appointment with the USDA Agricultural Research Service, when I was asked to look at the possibility of decomposing weed seeds and soil using microbiological approaches that no one had really looked at before. At the time that I came into the agency—in the early eighties—pretty much everything was chemical based as far as weed management and any sort of pest management. So to have a challenge like that was pretty unique at the time. I was looking for native microorganisms that could essentially rot seeds in soil. It was a pretty challenging task. That led into looking more at plant and soil microbial interactions and at natural ways of trying to control weeds using biological control approaches.

I started looking at what pesticides do when they're in a management system. That eventually led to my interest in soil quality, which was the big buzzword at the time. It evolved into soil health. With soil health, folks are looking at management assessment approaches. At the time that we really got into it, that's when the genetically modified crop issue came into vogue and into wide practice. On top of that, you add the whole Roundup/glyphosate issue. Certainly no one was really looking at that as being a factor that was affecting soil health. We diverged and looked into that for a while as well. So that hits about half a dozen of the highlights over the career that I've had. It always seemed like you turned the corner, went on to something else, and got another interesting challenge.

JOHN: In your early work in seeking to develop microbiological controls of weed seeds and weed germination, did you have any success in developing biological controls? Are there any tools available for farmers to use today?

ROBERT: A lot of the work that we started eventually served as a pattern for others to take up and pursue. My main task was to try to control these weeds in annual row cropping situations. It was almost impossible because—especially in our conventional input-based agroecosystems—so many weed seeds had built up over the years. Plus the fact that they were so diverse—you had a number of different weed seeds in the field. Although we found some bacteria and fungi that could affect maybe two or three weed species, in some situations, we would have seven or eight weed species in the field that would still be problematic.

In a midwestern situation, with corn and soybeans, it was almost impossible to come up with something that was effective biologically within the framework of the practices that were in place. I'm talking about a short rotation of corn and soybeans—no cover crops to build up the microbial diversity—and continuous chemical application.

But I will say that our work did set the stage for others in other parts of the country who have developed these types of organisms. For instance, in the Pacific Northwest, where cereal crops are the dominant production system and where there are only one or two weeds in the system—downy brome or jointed goat grass—they used this kind of approach. There are some organisms available today that are under development for commercial applications in those types of systems where it would work.

JOHN: That is fascinating. Are you aware of any development work that has been done that would be relevant for fruit and vegetable production systems, or where you have agricultural ecosystems that do have a longer crop rotation and that incorporate cover crops and so forth? Has there been any work done on those areas?

ROBERT: There's been a little bit. Agri-Food Canada has really supported this type of approach. For instance, there was a specific pathogen they developed for a particular weed in horticultural crops such as strawberries, and it was particularly effective. It was against a mallow weed—in the cotton family.

There are a few other examples, but the effort that has been put into these types of biocontrols—relative to the number of products that are available—is very lopsided. There's not that much out there. To develop such a product to go to market—there aren't a lot of companies out there that want to take that risk, especially if the crop is not high value. That's not going to bring them a lot of profit. That's always been an obstacle as well.

JOHN: One of the pieces that you mentioned was the impact of pesticides on biological processes in the soil. Could you elaborate a little bit on what you've observed the impact of various pesticides to be?

ROBERT: It's been somewhat variable. What we're finding is that some pesticides will be detrimental or toxic to just a group of microorganisms in the overall soil community. The same ones may stimulate other components of the microbial community. You end up with this imbalance of the various microorganisms that are in the soil.

A very good example of this was in the late seventies and early eighties, when we had a lot of soil-applied herbicides, fungicides, and insecticides that were going on at very high rates. You might recall Furadan as one of the insecticides and EPTC or Sutan herbicides. They went on in pounds per acre. The other interesting issue with those chemicals was that their chemical structure included a group that was very similar to what you find in proteins—the peptide bond; they call it the carbamate bond. These really stimulated a lot of microorganisms that were able to use proteins. They just adapted and used those for metabolism. The first thing you know, we had a lot of pesticide failures due to that.

Then we evolved into using more postemergent herbicides, and everyone thought, "Well, that's going to take care of that issue." And do you know what happened after that? We begin getting a lot of herbicide resistance. This was even before we had the Roundup-resistant weeds. We had a lot of ALS resistance and even atrazine resistance in the weeds.

What was also happening—at that same time that the continuous application of these herbicides was causing weed resistance—they were also altering the soil microbial community. We often forget—like in the case of glyphosate—that even though we're applying the herbicide to take out the weeds within a genetically modified crop that is tolerant to glyphosate, only about 5 percent of what's applied is being taken up by the weed. The rest of it is taken up by the resistant plant—the resistant crop—or is reaching the soils by some means.

Additionally, being a systemic herbicide, that compound is released throughout the plant, and it goes to the growing points. These include the roots, so you have a lot of that compound being released into the soil—into the rhizosphere—where you have the most active and dominant communities of microorganisms. That's where all the nutrients are being released by plants, and that's where you find most of your microorganisms.

Over the last twenty years, glyphosate is the one herbicide I've worked with most. We've documented that glyphosate will affect, or is active in, the root zone of genetically modified crops, just like it is in the root zones of weeds. It's released. It causes a lot of leakage of nutrients and of the sugars, carbohydrates, and amino acids from the roots. This stimulates a certain segment of the community—in this case, a lot of *Fusarium* fungi and some of the other detrimental bacteria, like agrobacterium and bacteria that oxidize manganese, making it unavailable to plants.

Often, the resistant crop—let's say its soybean or corn—will not be infected by these organisms because they're able to resist them. But they are setting up a shift in the microbes in the soil that can be detrimental to other plants. It can be detrimental to other microorganisms. And now, even though we've been told that glyphosate dissipates and is neutralized and is degraded, we're finding residual glyphosate in the soil now. We don't know exactly what that residual glyphosate—at whatever concentrations it's at after the crop is gone—what it's doing to the whole soil ecosystem.

JOHN: This is fascinating. How long do you find that glyphosate residues are remaining in the soil profile after application?

ROBERT: We started looking at this, at least in a cursory manner, beginning in 2014. We were looking in field plots at the time because one of our projects was determining whether we could come up with some treatment that might be able to overcome the effects of glyphosate on the soil—on the microbial activity and its function.

Half the plots had not received glyphosate for at least a year, and the other half had received glyphosate during the growing season. We found that some of the plots that had gone a year without Roundup application had as high residual glyphosate levels as the plots that had received the glyphosate in the current year. That was amazing to us because it showed that we were seeing a carryover of the glyphosate.

The other interesting thing was that it was random. You know how a field plot design is set up in a very structured arrangement. But one plot that received glyphosate might have over one thousand parts per billion, and then another plot that received the same treatment may only be at 350. So there are a lot of factors going on that affect persistence of this particular active ingredient.

I would just add one more thing. In the experiment I just described, we were only able to detect glyphosate. In the last two years, though, we've been using a different lab that is able to separate glyphosate from its major breakdown product, aminomethylphosphonic acid (AMPA). This metabolite can be just as toxic as glyphosate itself. So we had those two compounds analyzed—the glyphosate and the AMPA—and we found about ten to fifty times more AMPA than glyphosate. That indicates that glyphosate can be degraded by the microbes in one step. But the enzymes to further break down the AMPA are not there within that microbial community. They've been wiped out, either by intensive farming or by other herbicides or by glyphosate itself. It's pretty interesting. We're finding that AMPA could be carried over for more than a year or two as well.

JOHN: What are the implications of high concentrations of AMPA in the soil profile? Do they have a similar effect on the soil biology as what you described for glyphosate?

ROBERT: Oh yes, it has a very similar effect because the active group in glyphosate and in AMPA is the same. It's that phosphonate group. Listen to Dr. Don Huber talk about the chelation of micronutrients by glyphosate; AMPA will do the same thing because it has not been degraded to the point where it loses that characteristic. It can exchange with phosphorus in the soil, just like glyphosate.

JOHN: You mentioned that you had done some work looking at soil quality and soil health. What defines soil quality? What defines soil health? How do you know that you have a healthy soil?

ROBERT: I'll go with the term "soil health." "Soil health" is a little more holistic, and it really emphasizes soil biological function as well as the microbial component. "Soil quality" was originally proposed to see how good your management was in terms of providing acceptable agronomic production. It didn't really emphasize the environmental effects as much.

You have to look at soil health in the context of similar soils within a given landscape. There are several indicators that will suggest good soil health, but the two main ones are soil organic matter—or soil organic carbon—and microbial diversity. We're getting to the point where we can quantify differences in soil health based on what I call "soil organic carbon quality." Soil with high organic carbon quality has a fairly high proportion of its overall soil organic carbon or soil organic matter in a pool called "labile carbon" or "active carbon." That's important because that is the portion of carbon that supports the microbial community in the soil. Plus, it can serve as a source of substrates or energy for growing plants, as well as for the activity of the microorganisms that are involved in nutrient cycling and plant growth protection. Having a soil with a good proportion of labile carbon is really important.

Now, I can't tell you what that proportion should be. Maybe it's 25 percent or upward of 50 percent. It depends on the quality of organic matter that has been returned to the soil over the years. In an ecologically based farming system, where you have a lot of diverse plant material and maybe compost additions, active carbon is going to be quite high, and the quality is going to be high as well.

Connected with that is the microbial diversity. That primarily means that you have a diversity in types of fungi and types of bacteria, as well as a good component of things like protozoa, which complete the food web and feed on these fungal and bacterial communities to keep them in balance. These other components also recycle nitrogen and phosphorus through their decomposition of microbial biomass. Plus, the diversity bacteria is important because you need several different species in order to accomplish a function.

For example, if we have corn stover on the surface of the soil, or just under the surface, one microorganism cannot break that vegetative material down to mineralize the carbon and nitrogen out of it and form organic matter. You have to have a successional group of microorganisms. They're a consortia that work together to break all this material down and to get the beneficial effects out of it. We often see the effects of this lack of a diverse microbial community in soils that are not as healthy—soils that were planted to, let's say, corn that was genetically modified for Bt production and maybe Roundup ready for glyphosate resistance. Sometimes we find corn stalks in these fields three or four years after the crop was harvested just lying on the ground not decomposed. I believe what's happening is that we've caused an imbalance in the overall soil microbial community such that it just can't get the decomposition job done.

There are other things you can analyze to assess soil health, such as rate of respiration or carbon dioxide evolution out of the soil. These factors indicate good microbial activity and enzyme activity. There should also be some physical and chemical indicators of soil biology. These are fairly obvious things that we've already known about through soil testing, such as the amount of available phosphorus and pH, and then, on the other hand, the bulk density, the aggregate stability, or the proportion of aggregates in the soil. So we take all of those things together.

I don't want to get into the technicality of it, but we're coming to the point where we can put these indicators into the model and come up with a soil health index. That's going to become important as we begin looking at comparing management systems and as we use that information to come up with guidelines that farmers can use to actually improve their soil health. I think this is the most lacking piece of information right now—there aren't a whole lot of guidelines for the guy who's out in the field. We can research all this as much as we want, but it's not going to do any good until we have something that is usable by the farmer.

JOHN: If I'm a farmer today, how would you advise me to begin measuring some of these pieces? Are there any lab reports or any laboratories that are doing this type of analysis at the moment so I can begin establishing a baseline of where my soils are at?

ROBERT: There are a few around the country that will do this—both commercial and extension labs. There's a lab out in Nebraska called Ward Labs. They include a profile of your soil microbial communities that describes the different main components and gives you an idea of how much microbial biomass you have. Cornell University has a soil health lab that does similar things. And we have a Soil Health Assessment Center here at the University of Missouri, although it's still in the formative stages. It does chemical analysis well, and it does do one analysis for microbial diversity.

One can send soil samples to any of those labs. Some have reports that enable better interpretation than others. They're a good way to start assessing your soil health. Also, some states have incentives for farmers to do this. I know the Natural Resource and Conservation Service here in Missouri will subsidize some of the testing for farmers to have this done. I think NRCS (Natural Resources Conservation Service) nationwide has a soil health program to try to improve soil health and maintain it. They're trying to get farmers to get behind this and are paying them to do these assessments.

JOHN: Coming back to the conversation about glyphosate and AMPA, how does glyphosate—and the accumulation of glyphosate and AMPA over extended periods—impact overall soil health?

ROBERT: We know that there are some indirect effects of continuous use of glyphosate on soil health because we usually measure a lot of the biological parameters when we set up soil assessments. We see that there are some effects on some of the beneficial bacteria that are involved in plant growth promotion, such as producing plant growth regulators that stimulate root growth and other beneficial bacteria that will produce pathogen-suppressive compounds. We're noticing that glyphosate tends to suppress those beneficial groups of bacteria, so that has an effect on subsequent plant growth as well. So we feel that there's a problem there.

I briefly mentioned the effect on some of these microorganisms that are known to cause certain micronutrients to be immobilized and, therefore, not available for plant uptake. One of these is manganese. Manganese, of course, is very important for the activity of many enzymes that are involved in many of our metabolic pathways. If it's tied up, you may have poor photosynthesis. You may have poor amino acid formation because you don't have enough of it to satisfy the needs of the enzyme.

We've found, for example, that *Fusarium* that will colonize the roots of plants that are treated with glyphosate. *Fusarium* is a manganese oxidizer, so it will immobilize manganese. If it's on the root system, manganese is not going to be taken up. If it's built up in the soil—whether there's a genetically modified crop there or not—it's going to remain in the soil. It's going to continue to immobilize manganese. If you don't have a lot of available manganese, that's going to affect the overall soil health as well.

There are a lot of other things. Glyphosate may exchange with phosphorus in the soil, and then you have problems with either excess phosphorus, or if phosphorus isn't being taken up by the plants, it can become an environmental problem. We discussed the quality of the organic matter because we basically just use two crops as the source of the organic residues being returned to the soil. If we don't have the microbes there to decompose them, or if there's not a diverse enough quantity of organic substances to help build up soil organic matter, then that will affect soil health as well because—like I mentioned before—organic matter is one of the key indicators for good soil health.

JOHN: In our experience working with many different farms, I would say, just off the top of my head, that greater than 90 percent of the farms we work with have experienced severe manganese deficiencies. I wonder about the long-term effects of this. When you have glyphosate accumulation, you have this shift in the *Fusarium* population, and you have oxidizing organisms that are immobilizing manganese. What are the long-term implications of that manganese immobility in the soil profile? How long does it take before that manganese might be released and converted back into a form that the plants can actually utilize?

ROBERT: That is a concern. I think there are several issues here. You have the effect of the shift in the balance of the microbial diversity. It's shifted toward a lot of manganese oxidizers, causing manganese not to be available. I think that over time, if one were to alter the management to include, let's say, cover crops—or at least different crops within the rotation—this could stimulate other types of microorganisms that will help free that manganese or that will, at least, compete with those oxidizers to reduce their impact. So that would be one possibility. And how long that will take, it's hard to say. It may take a couple of seasons or maybe less. That's something that really needs to be looked into.

The other issue is something we mentioned previously: How much manganese can be immobilized or chelated by the residual glyphosate and the residual AMPA? That, I think, is a very serious issue—especially in soils where the texture is such, or the level of phosphorus is so low, that you don't have any competition with the glyphosate or AMPA. They will obviously chelate or immobilize manganese as well. I don't think we have any real good information on how long that can happen or what the extent of that situation is as far as tying up manganese over the long term. Taking all that together—the shift in the microbes, the residual glyphosate and AMPA, and the basically continuous corn-soybean rotation—if that continues, the manganese problem may persist.

JOHN: When you speak of bringing crops into a rotation to help reduce some of that manganese and to increase its availability, what are some crops that are really effective at having a reducing effect and shifting the biology and the availability of manganese in the soil profile?

ROBERT: I don't have any specific ones in mind, but certainly when you have a diversity of cover crops in a mix, there will be some that will support different microbial communities that are able to mobilize these micronutrients and others that can actually mobilize the nutrients themselves. A common example is the use of buckwheat or some of the brassica crops, which can mobilize phosphorus or neutralize nitrates.

Then let's say you add something like sorghum to the rotation—grain sorghum or sweet sorghum. From my experience, sorghum has the ability to host a lot of mycorrhizal fungi in its root system. And mycorrhizae are very adept at mobilizing many nutrients—not just phosphorus. If you could add a crop like that, or other crops that can host mycorrhizae, that would be a very good way to get around the manganese problem to improve regrowth and to improve the overall diversity of the microbial community.

JOHN: Earlier, you mentioned the impact of genetically modified plants themselves, apart from glyphosate and AMPA. How do GMOs impact the soil's microbial community?

ROBERT: There's not a lot of information. We found with soybean, for example, that genetic modification can have what are called "pleiotropic effects"—indirect effects due to the genetic modification that are in addition to the intended effect. In other words, effects that are in addition to the effect of making the plant resistant to glyphosate. There are things that can happen in the root system—with some of the early genetically modified soybean varieties, anyway—that even without being treated with glyphosate, the roots seemed to release a lot more carbohydrates or soluble carbon and amino acids. This is problematic because it attracts a lot of microbes that readily use this material, and many of those can be potential pathogens. You have a potential problem not only with some root pathology, but it's also possible to build up these segments of the microbial population and carry them over from year to year.

Another situation where we find these effects is in corn. Not in all varieties, but in many varieties that had been genetically modified to be resistant to insects using Bt, there was a side effect where some of the corn stalks would have a lot more lignin than others. Lignin is very difficult to decompose. That's one of the reasons we sometimes see a lot of that residue being carried over for two or three years in the field—there's so much lignin that it can't decompose very fast.

I think there are other situations that can occur. I had a Brazilian student here who looked at some of the nutrient composition. Some of the omega fatty acid ratios were changed in soybeans due to the genetic modification—that kind of thing. I can't say for sure if that has changed with some of the more recent cultivars because I haven't been looking at that very closely over the last few years. But as you know, in our commodity agriculture, these varieties change almost from year to year. Some of the varieties that we were using fifteen years ago are not available anymore. So that's always another problem. You just don't know whether the effects of these newer varieties are any better or any worse unless somebody has a research program that's addressing it.

JOHN: Your first point is very intriguing. In essence, what you're describing is that these crops and these plants may have the capacity to actually develop a disease-enhancing soil profile—which is interesting when you consider the long-term implications.

ROBERT: Right. That was a completely unexpected result that we had. We were comparing it to some of the old non-GMO varieties like Williams 82 and Maverick, and they had much lower soluble carbon and amino acid release. It was quite interesting, to say the least.

JOHN: What do you believe to be true about modern agriculture that many other people don't believe to be true?

ROBERT: There are a lot of issues. We can go back to this whole organic matter issue. We talked about the intense of use of chemical inputs in GM crops. We're seeing a depletion of soil organic matter content in these fields. I've been tracking some fields for thirty years or more. And soils that, at one time, had, let's say, 3 percent organic matter are now down to almost 1 percent. I think a lot of folks will not dismiss that like it isn't happening. When you use intense nitrogen inputs, like anhydrous ammonia, and you're not putting in more organic residues—whether from the crop, or with cover crops, or with organic amendments like manures and compost—that nitrogen has to be used. If it's not all used by the crop, the microbes are going to use it, and they are going to use whatever carbon is available in the soil. I'm certain that's where a lot of that depletion in the organic matter is happening.

There are some good studies from the University of Illinois that have shown this. And even they were very much criticized for it. People didn't want to believe it. Of course, a lot of it comes down to money. It comes down to the fertilizer companies trying to make money. If you tell them that some of their products are ruining some of the soil properties, that's not going to make them very happy.

JOHN: This idea of the importance of the soluble carbon component has come up in conversation a number of times recently, and you've just expressed how we are losing that. How can we regenerate that in the soil profile?

ROBERT: It all goes back to having a diversity of plants in the soil and having living roots in the soil as many months of the year as possible. When you can do that, and when you harvest as much carbon dioxide and sunlight as possible—to produce carbohydrates in the plant—and if you have a good profile that is inhabited by roots, you will get that soluble or labile carbon back. That's because, for all plants, anywhere from 20 percent to maybe 50 percent of the carbon that is fixed through photosynthesis will be released through roots by the plant. Not because the plant is compromised with a disease or anything—it just naturally does that. That carbon supports the microbial communities in the soil.

I think something as simple as mixing up that rotation helps. It's too bad that we don't grow as much wheat in the Midwest as we used to. That would be an excellent way of doing that. You follow corn or soybeans with wheat. You have wheat going into the winter. You can frost-seed clovers or lespedeza or whatever in with the wheat the next spring, and you have a nice cover crop going into the next planting season. What better way to get your soil revitalized than doing something like that?

JOHN: Farmers we have observed who are developing strong cover crop systems, such as Gabe Brown and Dave Brandt and others, are regenerating their soil carbon and their soil organic matter very rapidly and are producing some very nice crop results as well. I think there's a lot of potential.

ROBERT: Exactly.

JOHN: What is a book or resource that you often recommend to growers to learn about any of these various systems that we've talked about?

ROBERT: Well, there are some that are useful and that are not that technical. *Teaming with Microbes* by Jeff Lowenfels is good. It's a good overview of what's going on in the soil, and it describes what these different organisms are. I know there's another called *Teaming with Fungi*, but I don't know if it's much more informative than the first one.

Another one that I found very interesting is called *Mycorrhizal Planet* by Michael Phillips. You can get it from Acres U.S.A. It's really good. If you want to focus on the mycorrhizae and what they are, what they do, and how to establish them in the soil, I think that's about as good as it gets at the high-school level. Those are two books with a good microbial emphasis.

JOHN: What is the question you wish I would have asked?

ROBERT: I get asked to evaluate a lot of biological products. I don't know if a lot of the folks that you work with have looked into these kinds of things, but some of these products are very interesting, and some of them, I would not recommend at all.

JOHN: So of these various biological products, what are some of the ideas or the technologies that you're really excited about?

ROBERT: There are a lot of different classifications of biological products, and I don't think anybody has really settled on anything that is set in stone at this point. But there are things like biostimulants that help with the microbial community activity as well as the plant, especially in terms of root growth and plant growth regulation. There are things that are probiotic-type biological products, and these have the actual living organisms in them, and they are supposed to be selected to help promote growth of plants and maybe help some of the soil functions, such as cycling nitrogen and phosphorus and these kinds of things. Some claim they have nitrogen-fixing bacteria, but I haven't had much luck in showing that this actually happens.

Then there's another class of products known as prebiotics. There are quite a few of these out there. A prebiotic is a solution or a product that is made of the metabolites or the biochemical products of microorganisms, but they do not contain living microorganisms. They stimulate growth of the plant. I've seen the chromatographic spectrum of one product, and it shows that there could be 300 different compounds in this thing. I don't even know what they all are or what they all do. But I do know that some of them stimulate plant growth. Some of them are plant growth regulators like cytokinins and gibberellins, and some of them stimulate the microbes in the soil.

Here's the issue: most of these products—their effect or their efficacy is primarily based on testimonials. They haven't really been tested by unbiased parties. But there are a few out there that have been tested. We've tested some, and some have been tested by other institutions or companies to show some effect.

I guess what I'm saying is that those that I've looked at that work really well, I'm impressed with. Some of them have specific plant growth regulators that really help with root growth. They help with plant vigor. We find that they will also stimulate certain microbial groups in the root zones or the rhizospheres of these crops.

Of course, here's the other caveat to all this: a lot of these companies will approach us and say they want to test their chemical to see if they can offset the effects of glyphosate because they know there might be an angle there—it might make their product a little more valuable to the farmer if they can say, "Glyphosate has these detrimental effects, and it has these side effects, but our product can counteract that." In some cases, we've been able to show that to be true. We've tested some of these products for more than two or three or four years, and we consistently see that there is an offset.

But the biggest issue I've had is that some of these products will do all that, and they'll even actually help the soil health indicators and help plant growth. But then at the end of the day, at the end of the season, we don't see that much increase in yield. I think that's the bottom line with these products: if you're going to buy them, you want to have enough of a yield increase that it will pay for the product. A few years ago, when we had higher prices for corn, there were a lot of people using these things. Right now, not so much. But I do believe that some of them are worth the money because I've seen how they work. I think that they not only help the crop; I think they help offset some of the detrimental effects on the overall soil community due to some of the modern agricultural techniques that we're using.

JOHN: We use a number of different biostimulants, probiotics, and microbial inoculants in the crop programs that we put together. As we have tested many different products from different companies over the years—on a field scale—one of the biggest challenges we have observed is that we are often not able to replicate trial results. There's an inconsistency to the results. We might have a product that produces extraordinary results perhaps 30 or 40 percent of the time. It works really well sometimes, but it doesn't work consistently. That's been a very interesting observation because it's occurred with quite a number of products.

This has led me to wonder whether there is also an environmental component. Do we really have the right conditions in the soil to give an inoculant or a probiotic the opportunity for success?

ROBERT: I think you're right onto something there, John. That's what I tried to indicate to the manufacturers when we test these products because we're seeing the same deal. That's why I wouldn't make any conclusions on what I've seen on just one year of data. We have to have at least two or three years. Even at that, sometimes we do have inconsistent positive yields.

As far as the performance of these products, that point is well taken. I say that because we're located in a part of the state that doesn't have the best soils for producing corn and soybean. It has this claypan, and conditions have to be just perfect in order to get a very good or an acceptable harvest from year to year. So I think we're testing some of these products under soil conditions that are pretty rigorous. If we were testing in some of the mollisols or more fertile soils in the Upper Midwest, we might see something completely different.

These companies—I know that they are not as flush with cash as some of the big ag chemical companies, but if they really want to show how these products perform, they need to be testing in different regions of the country, or even internationally, in order to really show how these products perform.

JOHN: Absolutely. What is the one action that you would recommend growers to take?

ROBERT: One of the principles of sustainable agriculture is we have to keep the soil covered. That doesn't necessarily mean that we have to have cover crops on, but after harvest, the soil needs to be covered. If it's bare, or if there's not much cover with the residue, it is exposed to the elements. There's hardly any restoration of moisture in there—if it's there. I think that is a key.

Keeping living plants on that soil is even better yet because we get that organic matter built up—it will absorb more soil. You won't have a lot of erosion, and infiltration will be more regulated. So I think those are keys.

But I have to say one more thing. If we have a little time—and this is a practice that we used to do forty or fifty years ago—growers should seek to integrate livestock with the cropping system. I believe that if we could run livestock out on the fields—and I know it's not going to happen, except maybe for family farmers—it would help process the residues, and grazing cover crops can recycle some of the organic matter and stimulate the root growth by grazing. I think that's a great practice that we have pretty much forgotten about.

JOHN: That's a very interesting comment, Robert, because that's come up in a number of conversations recently—how livestock can have an impact on rebuilding soil organic matter and rebuilding the entire ecosystem to a much higher plateau, a much higher state of health and overall vitality and performance.

There's been some discussion about the economics of beef production. If we were to manage our livestock and our forages with the same level of intensity that we manage some of our other crops, the economics may actually be there to outcompete and to outperform current commodity crops, particularly with commodity crop prices such as corn prices being where they are right now. Do you have any insights to offer on that? What are the economic implications of managing grazing very closely and very carefully compared to current commodity crops?

ROBERT: That's a good question. I don't have a whole lot of insight into that, but your example of Gabe Brown is perfect for this. His system integrates grazing herds into forages, which are then rotated eventually into row crops, as I understand it. So the economics, I think, would be there. I believe that the more exposure livestock have to grazing—not in confinement—that meat quality is going to be so much better than what you find from a feeding floor. If you pushed the pencil on that, I believe you would find that the overall profits in the system would very much outweigh the way we are doing commodity crops currently.

To me, commodity crops are emphasized a little bit too much. They're for the commodity market, for exports, for biofuels, and all that. Very little of it goes to feed, although a lot of soybean meal goes to livestock feed. We've really gotten away from what an agroecosystem should be. I think that when the family farmer or the smaller-scale farmer had livestock integrated into the system, the profitability was there. It's just that it's been hijacked by industry. I think that's the bottom line. If there were a way to turn that ship around, I believe a lot of folks would be surprised how profitable that would be. But unfortunately, I can't give you any solid figures at this point.

WORKING *with* LIFE RATHER THAN DEATH

GABE BROWN

When I was in my conventional mindset, I used to wake up every day trying to decide what I was going to kill that day. Was it going to be a weed? Was it going to be a fungus? Was it going to be a pest? I was going to kill something every day. Now I wake up thinking about how to get more life onto my operation, and it's much more enjoyable working with life than with death.

We can't feed the world with the industrial mindset. We can produce enough commodities, but we can produce enough nutrient-dense food. We can do that with the regenerative mindset because we stack so many more enterprises per acre.

Too many producers look at their soils as simply a medium to hold the plant upright. They don't realize that soils are a living, dynamic, resilient ecosystem, and they don't realize the power of soil life—the power of photosynthesis.

I'm not a proponent of the current production model. I think it leaves too many decisions in hands other than our own. Our farm is profitable every year because we set our own prices.

WHAT IMPACT would four years in a row of crop failures have on your operation? For Gabe Brown, it was the best thing that could have happened to him. It forced him to adopt regenerative methods that have turned his operation into one of the most biologically diverse and successful ranches in the nation.

Gabe is a native of North Dakota, and with his wife and son, Paul, runs Brown's Ranch, a direct-to-consumer operation that sells everything from beef to honey to eggs. Gabe is a partner in the regenerative agriculture consulting company Understanding Ag and the educational firm Soil Health Academy. He is the author of the bestselling book *Dirt to Soil*, which tells his story of those four years of crop failures in the nineties and how they were the beginning of his successful journey.

JOHN: Gabe, you have a fascinating story. You've had many successes on your farm and have been able to make some remarkable changes over the years. What are some of the most memorable moments that led you to where you are today?

GABE: I was born and raised in town—I wasn't from a farm or ranch. But I took an interest in farming at a young age and happened to marry a girl whose parents had a farm. They were very conventional—heavy tillage and use of synthetics, etc.—but because I was not born and raised on a farm, I didn't have any of these preconceived notions. I was open to new ideas.

In 1991 my wife and I purchased part of the operation from our parents, and in 1993 a good friend of mine in northern North Dakota convinced me to try no-till. It didn't take a lot of convincing because it just made sense to me. In our rather semiarid environment, we're trying to save moisture. He said we needed to go no-till in order to save time and moisture. He also said to sell all of our tillage equipment so that we'd never be tempted to go back. I actually did that. I had to sell my tillage equipment in order to be able to afford the no-till drill we purchased, and so that started me down this path.

The biggest change came about between 1995 and 1998. We lost two crops in a row to hail, then one to drought, and then the fourth to hail again. I tell people that those four years of natural disasters were absolute hell to go through, but they were absolutely the best thing that could have happened to me. The bank wasn't going to loan us any more money, so I had to learn how to make a living off of this land without the use of synthetic inputs.

Those were the biggest driving forces in our move toward regenerative agriculture.

JOHN: What were the key pieces that you changed during and after that four-year period that were substantially different from the way that you had been farming before?

GABE: The first year—1995, when we lost 100 percent of our crop to hail—the day before that storm, I was going to start combining our spring wheat crop, which was our major cash crop at a time. All of a sudden, the hail had laid all of this biomass down on the soil surface, so we got all this armor protecting the soil. Then in 1996 the same thing happened—the hail laid down a tremendous amount of biomass and left the soil intact. We weren't disturbing the soil ecosystem. Because that hailstorm was so devastating, we didn't even have enough feed for the livestock. I had to go in and seed sorghum-sudangrass and cow peas, and I was intelligent enough to know the value of seeding legumes with grasses and other simple things we can do to benefit production and increase biomass. It got to the point where, with virtually no money, I couldn't even afford to put up hay. We turned livestock into those acres and grazed them.

This series of natural disasters was actually the best thing that could have happened from a soil ecosystem standpoint. We were lowering synthetic inputs, which obviously led to not as much carbon being burned up due to the overuse of nitrogen fertilizer. We weren't supplying phosphorus fertilizers, which led to the build-up of mycorrhizal fungi because of the lack of soil disturbance. All these little things that nature was forcing us to do actually advanced soil function.

I tell people that I very honestly had no idea what a cover crop was back then. I had no idea about soil health or soil function. Due to the circumstances, though, we were going down that path.

JOHN: Many people are beginning to recognize the value and the importance of cover crops, but there's still a lot of hesitation around incorporating livestock into agricultural ecosystems. In your estimation, how important do you believe it is to incorporate livestock to the overall ecosystem?

GABE: Let's step back and take a look at how our soils were formed. They were formed over long periods of time by large herds of elk and bison being moved across the landscape by predators—trampling, grazing, and then defecating and urinating. A natural nutrient cycling occurred. The animals moved on, so there were long periods of rest and recovery, which allowed for a maximum amount of carbon to be pumped into the soil. Soil health revolves around a carbon cycle.

That's how our prairie soils were formed, but look what we've done with our modern monoculture-type mindset. We've removed the diversity, and we've removed the animals from the landscape, and now we're simply not pumping as much carbon into the soil. Now, can we advance soil health without livestock? Absolutely. I have several quarters of land that we're just not able to graze livestock on because they're surrounded by housing developments, and there's no water. It's just not worth the hassle to incorporate livestock onto those acres. We've still significantly advanced the soil health on those acres, but those soils will never reach the health of the soils where we are able to graze livestock.

There are many benefits of having livestock on cropland. The one thing I think producers are really missing out on is the profitability of doing so. I ran some numbers after 2016, and we added $220 net return per acre on the cropland acres where we were able to graze livestock. With margins where they are in commodity agriculture today, that's significant. What producer doesn't want $220 more per acre? I'm not saying all producers will want to do the things we're doing with livestock; I'm just saying that they're missing a major opportunity there.

Cover crops are an absolute no-brainer from a soil health standpoint. If you can integrate livestock, that's a way to convert those cover crops into dollars. I tell producers that even if they don't want to do it, there are many people in their communities around the country who would love the opportunity to run some livestock—to enter agriculture that way. Why not benefit your land, help a young person, and integrate some livestock under your cropland?

JOHN: That's a great idea for including the next generation. When you talk about the economics and the profitability, an additional $220 net per acre—isn't that a larger net than most growers are getting producing corn?

GABE: It is. On our operation, we sell very few commodities. We're doing value-added, direct-to-consumer products simply because of the dollar return we can generate by doing so. I just shake my head at commodity agriculture today. I have no desire to produce commodities, and I really don't understand why people do it. There's just no money in it unless there's a major drought somewhere or some external force. We tie our hands, so to speak, when we produce commodities.

I'm not a proponent of the current production model. I think it leaves too many decisions in hands other than our own. Our farm is profitable every year because we set our own prices. We take that out of somebody else's hands and put it in our own. When you're able to do that, that really adds to your bottom line and increases profitability significantly.

JOHN: There is this entire industrial ecosystem that has evolved to handle commodities—all of the combines and equipment manufacturers and grain elevators and feedlots, which have all evolved to accommodate this commodity cropping mindset. What you're essentially describing has the potential, in my understanding, to displace practically all of that by bringing livestock back onto the land to graze crops and cover crops directly on the soil. That has some substantial ramifications—not only for long-term soil health, on a macro level, but also for the agriculture industry as it exists today.

GABE: Yes, you're right. But let's be honest—it's not going to change overnight. It's going to change slowly over time.

I spend a lot of time dealing with the human health aspect of production agriculture, and we have a human health crisis in this country. America spends more on healthcare than any other country in the world, yet we lead the developed nations in cancer, ADD, ADHD, Parkinson's, Alzheimer's, autoimmune diseases, obesity—the list goes on and on. Why is that? I'm in no way blaming it all on farmers and ranchers. It's due to our sedentary lifestyle. But farmers and ranchers have to accept part of the blame because the nutrient density of the foods that we're producing is a fraction of what it was fifty to seventy years ago.

How are we going to change this downward trend in human health? It has to start with the soil, and it has to start with the movement of nutrients through the ecosystem, which inevitably end up in our bodies. What's causing the downward trend in nutrient density in our foods? A lot of it is related to the health of our soil ecosystem. That, I profess, comes about from our industrialized, commoditized mindset. If we're going to truly affect human health, it has to start at the soil level and the ecosystem level and in what farmers and ranchers are doing. Then it gets back to using the principles of a healthy soil ecosystem. We have to do what's necessary to advance the health of our soils to get more nutrients—the plant secondary metabolites that are moving through the plant—into the human food chain. We're not going to do that with commodity agriculture.

I've spent a great deal of time with several large organic companies working on regenerative protocols. They are actually driving change on their producers, and their producers are going to be paid accordingly for the regenerative practices they use on their operations. It all comes about because these large companies know that the next real wave that's coming, so to speak, is the focus on human health. They're going to make more money if their foods have higher nutrient density. The only way they can get that is by improving the health of the soil ecosystem.

Farmers and ranchers get it beat into them—"We've got to feed the world, feed the world, feed the world!" Some figures released in December of 2017 stated that we produced enough food on this planet in 2016 to feed fourteen billion people. Think about that: we have approximately 7.8 billion people in the world, but we produced enough food to feed fourteen billion. Farmers and ranchers are shooting themselves in the foot if they think they have to increase production to feed the world. There's already enough production out there.

The only way you're going to differentiate yourself from everyone else and increase the value of what you're producing on your operation is if you produce nutrient-dense food. Then you have to set your own prices and be paid for it. That's what we're trying to do on our operation. That's what we're encouraging others to do also.

JOHN: This essentially indicates that we don't really have a production problem—we have a distribution problem.

GABE: That is exactly right, John. There are over a billion people worldwide who go to bed hungry every night, who don't have enough food, but it is exactly as you said—it's a distribution problem. It's not a production problem.

The other thing we have to be very honest about is that a lot of what we produce does not go into human food consumption. A lot of it is going toward things like ethanol and grain to feed cattle. I've slaughtered thousands of beef animals, and I have yet to see one with a gizzard, so why are we feeding these animals grain? It just does not make sense. In saying that, I'm going to be very honest and tell you that I will eat grain-fed beef if grass-fed isn't available. It's still a highly nutritious, nutrient-dense product, but I would certainly rather eat grass-finished beef, simply because of the health benefits of it. Plus, I prefer the taste of it.

Our soil health consulting partnership did a school in eastern Iowa, and it absolutely amazes me that the mindset of producers there is that the only thing they can produce is corn and beans. There's not the infrastructure or the market for other crops. I'm disappointed by that because that shows you how narrow a mindset we have in production agriculture. We've closed our thought processes to other crops that might make us more money.

I think part of the problem is the current farm program. Let's face it: 95-plus percent of cropping decisions are based on revenue insurance. I quit being involved in any farm programs a number of years ago because, for one thing, I don't think the American public should be paying and subsidizing my operation. If I can't make it on my own, I shouldn't be doing it. The other thing is, Why would I want to tie my hands and be limited to what I can do on my operation?

For instance, here near Bismarck, North Dakota, we've gone through a couple very dry years, and when we went into planting last spring, I had planned on putting more corn in. But because we were so dry, I switched my cropping decisions accordingly. I was able to harvest crops that made me a decent profit last year, even though we only had 5.8 inches of rainfall during the growing season. You tie your hands and limit yourself when you're in this commoditized, industrialized mindset. Producers need to learn how to adapt and observe, and to change according to conditions.

JOHN: In the state of Iowa, any crop that is not corn or soybeans is classified as a specialty crop—meaning alfalfa and wheat production are classified as specialty crops. Prior to 1946, I believe, Iowa was sixth in the nation for apple production. That climate still exists today. The climate hasn't shifted that substantially. There was one freeze in which they lost a lot of trees, and the entire industry never came back. I'm using apples as one example, but there was a diverse agricultural ecosystem in Iowa in the forties, fifties, and sixties. Northcentral Iowa was the flax production capitol of the world, and that has completely disappeared.

GABE: It's not just eastern Iowa. I can go anywhere in the United States and find similar stories, but the average net returns over the past nine years on my operation—and my highest net returning crop per acre of my cash crops—is cereal rye and hairy vetch combined together. Over the past nine years, our average net return per acre is $951. Now, how many people out there grow rye and hairy vetch and then combine it and sell it? Not very many. Bingo. People laugh at me because I'm different, but I laugh at them because they're all the same!

My family and I have planted over 1,500 fruit and nut trees the past few years. How many peaches are being grown in North Dakota? You know what I can get for a North Dakota–grown peach? Two to three dollars per peach. Why wouldn't I do that?

Here in the Midwest, we're somewhat isolated from what's happening on the coasts. In my traveling and speaking, I spend a lot of time on the East and West Coasts, and when you get to talking to consumers out there, it's absolutely amazing. I visited with a producer who is running eight thousand laying hens on pasture. He's selling his eggs for one dollar per egg, and he just laughs. He says that people can go to Walmart or wherever they want and buy cheap eggs, but he can sell all he wants at one dollar an egg because people are willing to pay for it. Here in Bismarck, North Dakota, we run 1,100 laying hens on pasture. You can buy eggs at the local store for less than seventy cents a dozen. We're getting $4.50 a dozen, and we can't even begin to keep up with demand because people see the difference when they crack those eggs open. They want to know where their food comes from. Farmers say, "I'm not going to raise chickens—I don't want anything to do with that." Okay, don't complain to me when you're not making money. This isn't rocket science.

Then think of how we're feeding those hens. I mentioned that we grow rye and hairy vetch that I'm selling as seed. We run it through a cleaner, take out all the screenings, and that's how we're feeding our chickens—just screenings. I'm taking a waste stream from one enterprise to fuel the profit in another. That's just good business. But so many people aren't willing to do these little things, and then they wonder why they can't make any money.

JOHN: I have many colleagues and friends who are doing similar things to what you're describing and are extraordinarily profitable. Not only are they profitable—they are finding a lot of joy and fun in the work they're doing, which didn't exist before they stepped outside of the typical boundaries of what everyone else was doing.

GABE: You're exactly right. I've had people in their sixties come up to me and say, "Gabe, agriculture is fun again. I'm enjoying this. We don't have this stress. My wife is happy—we're not stressed out over finances anymore."

When I was in my conventional mindset, I used to wake up every day trying to decide what I was going to kill that day. Was it going to be a weed? Was it going to be a fungus? Was it going to be a pest? I was going to kill something every day. Now I wake up thinking about how to get more life onto my operation, and it's much more enjoyable working with life than with death.

JOHN: You're exactly right. In a series of seminars a couple years ago, I asked farmers why they farmed—what it was that attracted them to farming in the first place. When you begin unwrapping the answers and digging down into the foundational things that attract people to farming—and to stay in farming—it really is their appreciation and desire to have a connection to life and to living processes. Farmers really love watching seeds germinate in the spring. They love watching the joy of a newborn animal. But we have unconsciously adopted a model of agriculture that is directly antagonistic to those core values. We've adopted a model that's based on a warfare mentality of search and destroy: identify a specific pest or pathogen and figure out how to kill it. And then, of course, if you aren't successful with the first weapon of choice, just get a bigger bomb—a stronger pesticide.

GABE: That's right, and look where that's led us. We now have weeds that are resistant to herbicides, and it's leading us down this path of decimation of the ecosystem. It also has human health ramifications, and I really think that we, as producers, need to step back and look in the mirror and realize that the things we're doing on our farms and ranches are impacting human health. We have to take some responsibility for those actions.

JOHN: I think that in many cases, we, as growers, may have lost sight of the fact that our primary responsibility is growing food. We talk about feeding the world, but it's become a very abstract concept for many of us. It's no longer actually about connecting to the people who are consuming that food. It's easy to distance ourselves from that.

I'd like to come back to beef production. There is a great deal of dogma and a great deal of political dialogue around how beef is bad. Beef is bad for the environment. We are consuming too much beef. We need to reduce our meat consumption—we're not going to be able to feed the world if we continue to increase meat consumption as emerging and developing countries increase how much they eat. And yet this is a very different perspective from the one that you've been describing—you believe that livestock are actually an integral part of a regenerative agriculture ecosystem.

If you look at protein production per acre from a macro perspective—and being able to feed a growing world population a higher-quality diet—can we increase meat consumption globally and produce an equivalent or even greater amount of beef using the type of system you are describing than we can with corn and grain production?

GABE: That's a very good and pertinent question. The answer is absolutely yes. We can increase meat protein consumption. It's not only beef; it's other species of livestock also.

Look at what we're doing now in production agriculture. Animals are confined in lots—a feedlot, a large production dairy, a hog production building, a poultry production building—instead of having those animals out on the landscape. You often hear people demonizing beef consumption because of methane production, but what they fail to take into account is that there were actually more ruminants in North America before European settlement than there are now. If methane were a real issue, we should have had methane emission problems back then, right? The reason we didn't is because there are methanotrophs. When a grazing animal emits methane, those methanotrophs latch onto it and use it as a food source. That's not happening with animals in confinement.

The critics are correct when considering today's production model, but there are many benefits to getting animals out onto the landscape. It's a proven fact that the microbiology in the soil is directly related to biology in the rumen, and to the biology in the human gut. We have billions of bacteria and fungi in our guts. It's all tied together.

Let's look at the average producers who are growing corn and soy. They're producing one commodity each year—either corn or soybeans. Now look at a regenerative model. We can still grow corn and soybeans, but besides that, we're growing cover crops, and then we're grazing those cover crops with beef or hogs or poultry or lambs. I'm stacking so many more enterprises onto my land. I just laugh at the notion that we're not going to feed the world with regenerative agriculture. I'm producing way more nutrient-dense calories per acre than the industrial mindset is. Plus, I produce a wider variety.

We can't feed the world with the industrial mindset. We can produce enough commodities, but we can produce enough nutrient-dense food. We can do that with the regenerative mindset because we stack so many more enterprises per acre. Therein lies the big difference that nobody takes into account. They only look at monocultures because they have the mindset that we have to look at one single thing. We have to look at the ecosystem as a whole and all these stacked enterprises. We're going to produce way more nutrient-dense calories per acre in a regenerative manner.

JOHN: That's incredible. That's exciting. That's exactly what I've been observing with growers that we have worked with. That gives us a great deal of hope for the future, when we think about the opportunities that exist to develop an agricultural ecosystem that not only regenerates soil health and plant health but that can also begin a legitimate conversation about growing food as medicine—about helping to improve the health of consumers.

GABE: There's a book called *What's Making Our Children Sick?* that was written by two doctors. They take a look at what's happened to the health of society, particularly in the United States, over the past fifty to seventy years. The real decline in human health came about when we changed the production model from one that was very diverse to this industrialized mindset. As I mentioned earlier, I think a lot of it is tied to the health and function of our ecosystems. It's to everyone's benefit—the producer's, the farmer's, the rancher's—when we diversify and start stacking more enterprises because that's going to increase profitability—both for them and, inevitably, for society as a whole.

We haven't even started talking about healing our watersheds. Water is becoming a larger and larger issue throughout the world. Those ecosystem processes will also be healed as we change from the current production model.

JOHN: Gabe, as you have developed these systems on your farm over the decades, what has been something that has really puzzled you?

GABE: We're blessed—we get thousands of visitors coming to our ranch. It's amazing! They never come during the winter to North Dakota; it's always during the summer. We've had visitors from twenty-two foreign countries and all fifty states over the past five years, and every Canadian province, and they come thinking that it's going to be this utopia. Well, I've got a working farm and ranch like everybody else, and we have issues that we deal with, and one of the big ones that I've struggled with is perennial weeds.

As we moved from a conventional cropping system to a more regenerative model, the annual weeds ceased to be an issue. They just don't bother us much. But it's a puzzle to figure out the perennial weeds. I really believe that the work that Dr. David Johnson is doing at New Mexico State University, with a high fungal component in his soils, is going to be one of the keys to figuring out that puzzle. I have been a big proponent for many, many years of getting away from bacterially dominant soils. We have to get the fungal activity back into our soils. Dr. Johnson has helped me realize that it's even more important than I thought. I'm happy that our consulting business was able to hire Dr. Johnson as a technical adviser, and he's teaching us how what he's doing with static compost, incorporating high fungal activity, and it's really showing some promise in being able to mitigate these perennial weed issues.

That's one of the things that's puzzled me the most. Like everybody else, I'm very cognizant of the fact that I am farming a degraded resource. I've been on thousands of operations all over the world, and I've never been on a single operation—including my own—where the resource wasn't degraded. There's something for all of us to learn. It's like a big jigsaw puzzle. We're learning all the time, trying to fit the pieces together.

JOHN: Thinking about developing fungally dominated soils, you have a substantial advantage in that you've developed a system that incorporates a lot of perennials. You have an environment that can foster the development and the preservation of fungal communities in the soil profile. This is the critical element that may be missing on a conventional farming operation.

GABE: That's right. Many people tell me that their soils are different and that their climates are different. Yes, soil and climate are different everywhere. But the principles of healthy ecosystems are the same no matter where you are.

The other thing producers tend to not understand is that they can influence how those ecosystems function by their management. Every ecosystem, all over the world, has its unfair advantage. You mentioned some of the unfair advantages I have, being located in North Dakota. I just did a school in North Carolina. They have unfair advantages in that their growing season is much longer than I than mine, so they can cycle more carbon than I can. Now, due to their moisture, they're going to cycle through that carbon faster, but they can grow more things, so they can compensate for that. No matter where we are in the world, there are unfair advantages, and it's up to us as producers to find those unfair advantages and use them to our benefit.

JOHN: Exactly. Gabe, what has been something that has surprised you?

GABE: In 2003 I was very blessed that Dr. Kris Nichols came to work at Mandan, North Dakota, about twenty miles from my place. She came out to my ranch and saw what I was doing, and she said, "Gabe, you've come a long way. Your soils have improved significantly. But you will never truly be regenerative until you significantly back off of or eliminate the use of synthetics on your operation."

I'm a pretty gullible person in that if a person really comes across as credible to me, I am going to try to either prove them right or wrong. After Dr. Nichols was here, for four years, I did split trials. Our soils had advanced quite a bit by that time, but I did split trials with different rates of synthetic fertilizers, all the way down to zero. I was amazed that the production of cash crops I was able to get—and cover crops, for that matter—without the use of those synthetics.

The other thing that really struck me was that on the fields where we were not using synthetics, the aggregation of the soil advanced very rapidly. We noticed that we were building more soil aggregates. Of course, when you build more soil aggregates, you have better water infiltration, and you have a home for biology.

That really surprised me—that I was actually decimating my soils by the use of all these synthetics. I owe a great deal to Dr. Kris Nichols for challenging me back then. Our soils took a giant leap forward. We tried those split trials for four years, and we haven't used any synthetic fertility or pesticides or fungicides since 2007. We still occasionally use an herbicide, but not nearly as often.

JOHN: That is quite incredible. Just for clarity, when you talk about synthetics, particularly synthetic fertilizers, what had been the synthetics that you were using? Are you talking about things such as anhydrous ammonia or potassium chloride or some others?

GABE: We were not using any anhydrous ammonia. We were not using any liquid forms; it was all dry forms of synthetic fertilizers—urea and DAP, etc.—simply because our equipment was set up for that. We noticed that we were significantly overapplying nitrogen, and of course that was burning through the carbon; it was destroying aggregation. The other thing was that we were overapplying phosphorus. This is something that is done virtually all over the world, and it has a negative impact on mycorrhizal fungi. Mycorrhizal fungi, of course, produce glomalin, which is critical for the formation of soil particles. That combination is why we noticed such a significant improvement in the aggregation of our soils.

Now am I sitting here today telling producers that they should all quit using synthetics? No, don't do that—you're going to have a wreck if you do. Our soils are like drug addicts: we have to wean them off of these synthetics over time. The beautiful thing today is that we're able to do that, thanks to tests such as the Haney soil test, which gives producers the confidence to see how many nutrients in their soils can be converted from inorganic to organic via biology. There are huge advancements being made in soil testing and in how nutrients move through the system via biology. We're able to more confidently start weaning producers off of these synthetics slowly, over time, in order to get the health back into their soils.

JOHN: That's one piece that we tell growers in our consulting work: they need to earn the right to discontinue the use of fertilizers and also to discontinue the use of fungicides and insecticides. There's tremendous potential to grow plants that are extremely disease and insect resistant, but you have to earn the right to get there, and it doesn't happen overnight.

GABE: That's very well said and very true. Look at the work of Dr. Jonathan Lundgren—he just released a peer-reviewed paper that shows that regenerative farms and ranches are much more profitable, and they simultaneously improve their ecosystems compared to those that are using, in this case, stacked-trait corn genetics.[3]

JOHN: Exactly. Gabe, what do you believe to be true about modern agriculture that many other people don't believe to be true?

GABE: Today, when producers go out and buy a new combine or a tractor or a new pickup, they get an owner's manual, right? We're never given that for soils. When we go out and buy or rent land, we're not given an owner's manual.

Too many producers look at their soils as simply a medium to hold the plant upright. They don't realize that soils are a living, dynamic, resilient ecosystem, and they don't realize the power of soil life—the power of photosynthesis.

I'm often asked by people who come to my place, or when I speak at conferences, "What's the one thing you do on your operation that sets you apart from your neighbors?" I tell them that I have a living, growing root in the ground as long as possible throughout the year.

Last October, I had to drive 652 miles to speak in Western Montana. How many green, growing crops did I see in October from Bismarck, North Dakota, to Butte, Montana? Once I left my place, the answer was one. I saw one field that had green, growing crops on it—that was taking CO_2 out of

---

3 Claire E. LaCanne and Jonathan Lundgren, "Regenerative Agriculture: Merging Farming and Natural Resource Conservation Profitably," *PeerJ* 6:e4428 (February 2018), https://doi.org/10.7717/peerj.442826.

the atmosphere through photosynthesis, converting it to liquid carbon, and pumping it into the soil.

We as producers do not realize the power of the natural system. If we would just give nature half a chance, we'd find that she's very resilient. We could heal our soils, put more dollars in our pockets, and produce more nutrient-dense foods. That's what frustrates me. Too many producers have blinders on—they think it's all about federal farm programs. We've made ourselves resilient to trade wars because of what we've done to heal our ecosystem. Producers don't realize just how much of their profitability revolves around this living, dynamic ecosystem and that they can heal it if they give nature half a chance.

JOHN: That is one piece that I'm so excited about and that I speak a lot about at conferences: harnessing the power of the photosynthetic engine. Having living, growing plants is really the only way that you are bringing new energy into the system and into the soil profile, and it's so incredibly powerful. Most growers have never had the opportunity to observe truly healthy plants that are photosynthesizing at max capacity. If we could just harness that photosynthetic engine and tap into it, this would have such incredible possibilities for regenerating soil health and producing extraordinary crops. There's tremendous untapped potential there for everyone.

GABE: You hit the nail on the head there, John. Ray Archuleta always poses this question in his presentations: Is your farm or ranch being run on ancient sunlight or new sunlight? Are you using high amounts of carbon in the form of fossil fuels, or are you harvesting energy with growing plants? Like Ray always says, if we could just get producers to cover their fields and grow things, and then to significantly back off on tillage—so carbon can cycle through rather than being released from a tillage pass—we could take care of a lot of the issues that we're seeing in production agriculture today.

This isn't rocket science. It's been done for eons of time. When livestock graze one of these cover crops—a living plant—that plant is going to start pumping more carbon into the soil, thus improving the carbon cycle even more. That's where livestock integration can advance soils to another level.

JOHN: Absolutely. What do you think is the biggest opportunity in agriculture today?

GABE: The biggest opportunity I see in production agriculture revolves around advancing soil health to the point where we can truly produce more nutrient-dense foods. We're seeing a real trend worldwide, but particularly in this country—where consumers are starting to think of food as health. Look at the amount of prescription medicines that are being used today. There is a shift occurring in consumers. Instead of doing that, consumers are starting to realize that food is health and that what they eat directly dictates how healthy they are.

I had the good fortune of spending a day with Dr. Daphne Miller. She's a family practitioner who wrote the book *Farmacology*, and she told me she firmly believes that approximately 95 percent of illnesses people have can be cured or prevented just through what people eat. As a producer, why would I not want to take advantages of that? Why would I not want to offer products that are high in nutrient density, that I'm going to be rewarded for? I'm a capitalist. I make no excuses for what I'm charging because I'm the one out there putting my capital on the line every year, using these practices that are necessary to ensure that what I'm producing is nutrient dense. Consumers have already proven to me that they're willing to pay. They'll write the check for it because they know that my food will have a positive impact on their health.

I think that's the one thing that more producers need to look at. I really think we're going to see that trend accelerate rapidly in the near future. Consumers are going to search out the producers who are producing nutrient-dense products using regenerative practices, and they're going to vote with their buying dollars—where and who they want to purchase from.

JOHN: I see that happening already as well, particularly in fruit and vegetable production, and it's just going to continue to accelerate. I think you're absolutely correct.

What is a book or resource that you recommend most often to growers?

GABE: There are two that I'm really recommending right now. One is *A Soil Owner's Manual* by John Stika. John is a retired NRCS soil health specialist here in North Dakota. It's a short, easy read that outlines the principles of soil health, and I hand out copies wherever I go because it's a way that producers can sit down and get the ideas in their mind of what they need to do to regenerate their resource.

The next book I've been recommending lately is the one I mentioned earlier: *What's Making Our Children Sick?* I do that kind of as a shock factor because, as I said earlier, I don't think we as producers realize the impact we're having on society by what we're doing. I think it's up to all of us to look in the mirror and realize that what we're doing may not be what's best—not only for ourselves but for society.

JOHN: What technology and ideas are you excited about for the future?

GABE: If people get to know me, they'll realize that I'm not a very intelligent person. I am observant, though, and I'm able to look at things. As I said earlier, I'm on hundreds of operations every year all over North America, and I love going to people's farms and ranches. I can always learn something everywhere I go.

I think we've lost the power of observation. I think sometimes we look at technology as a cure-all. It seems like we as producers want a recipe. We want to know that if we apply this pesticide or this seed treatment, then we can combine X amount. Regenerative ag is all about observation and about understanding what is happening in our soils. It amazes me how many producers never take a shovel and look at their soil. I've dug in several of my neighbors' soils more often than they have. They just never do it. They don't think of looking at it.

You asked what technology I'm excited about. There are leaps and bounds being made in the knowledge of soil biology right now—the instrumentation to analyze what this biology is doing. Dr. Kris Nichols, who's one of the world's foremost soil microbiologists, will tell you she used to think she knew 1 percent of what was going on in the soil. Now she realizes she knows one tenth of 1 percent. That shows you the amount of learning we have yet to do.

Another thing I am really excited about from a consumer standpoint is that the Bionutrient Food Association is working on an instrument that consumers will be able to use in any grocery store or farmers' market to scan a food item and determine the nutrient density of that item. I think that will be a game changer because consumers, if given a choice, are going to purchase the product that's more nutrient dense, if it's within a reasonable price. As consumers are more educated about food as health, I think that'll drive real change on the landscape, too, because all of a sudden, the products of some farmers won't be selling well. If they're way lower in nutrient density, they better start making changes to their management in order to positively affect soil ecosystems. I think that'll be huge when that comes out.

The third thing as far as technology is that there have been some real advancements in being able to measure exactly how much carbon we're cycling through our system. I'm working with a group called Landstream. Dr. John Norman, who helped develop much of the instrumentation that's been used by NASA, is a key player in that organization, along with Abe Collins. They're looking at real-time data—how much carbon—sunlight—is being collected over the landscape. They're going to be able to project the growth of plant biomass. This will really help grazers; they'll be able to predict how much forage they'll have available at any given time. Grain producers will be able to estimate how much carbon is cycling in their system and how much growth they'll get out of their plants.

JOHN: Gabe, throughout this conversation you've mentioned so many different ideas that farmers should consider. What are the first actions that you would recommend growers take?

GABE: That's a great question. You're absolutely right. A lot of times, people are just overwhelmed by it. It all comes back to the five principles.

The first thing I tell producers is that they need to educate themselves. With internet technology today, it's very easy to educate yourself—through YouTube videos, through Audible books to listen to while you're in the tractor. Look at that as school and learn something. One of the things that I'm fortunate to have is this desire to learn as much as I can. I try to read, on average, two books a week, and listen to more than that. I just can't get enough information.

I encourage producers to learn and then to apply the five principles. I tell them to take one field—whatever will allow them to sleep at night, whether it be five acres or a thousand—and move that land into regenerative practices. Try the principles. The first principle is simply to have as little mechanical or chemical disturbance as possible. The second is to always have armor on the soil surface—we don't want bare soil that doesn't have anything collecting sunlight. The third principle is diversity. We need diversity. These ecosystems were very diverse before we started farming them. My son taught rangeland management at the local community college for five years, and he brought his students out to one of our native pastures. In a two-hour time frame, they collected over 140 different species of grasses, forbs, and legumes. That's diversity. Now what are we doing today? Oh, we plant corn and beans. Which is going to be a healthier soil ecosystem? The fourth principle is having a living root in the ground as long as possible throughout the year, collecting sunlight. The fifth principle is animal integration, and I would add insects in there—little insects are animals too. We don't realize the benefit that insects provide to the ecosystem.

I encourage producers to try these five principles on a given field for five years. I've never had anyone go all five years and then go back to the old way they were doing things. I've had them drop out after a year or two because they just didn't have the will and fortitude to stick with it, but I've never had them go all five years. Usually by year three they're changing their entire operation.

There's one other advancement I just want to mention briefly here: we need to get perennials back into our systems in one way or another. Those who live a bit further south than you or I should consider what Collin Seis in Australia terms "no-till cropping," where you have a perennial growing at all times, and then you're actually no-tilling your cash crop into that. It's being done successfully in Kansas. Producers such as Shane New are doing this now—he has cool-season perennial paddocks, and then he's no-tilling warm-season species in there, such as grain sorghum. He has that living root in the soil all the time, and then as that cool season shuts down, he seeds the warm-season cash crop. It's the best of both worlds. I've had the good fortune of being at Collin Seis's place in Australia several times; to me that's the ultimate cropping system.

JOHN: That sounds like the next generation for certain. Gabe, what is the one question you wish I would have asked?

GABE: I understand that it's a bit overwhelming, but I tell producers to just try something. Just do something, but don't think you have to do it all. I don't want everybody doing all the things I'm doing. We all have our own desires, our own goals, our different landscapes and different conditions. You need to do what makes you happy.

But don't ever complain to me that there's not money in production agriculture. I often ask groups of producers the question "How many of you want your sons or daughters to come back to join the operation?" Most people raise their hands, and then I ask, "Well, how many of them do?" Very few do, and I think it's sad that we have an industry where we're actually driving young people away rather than inviting them in. My wife and I have two children, and our thirty-year-old son is a partner in the operation. I've known since he was young that he wanted to come back, but I made it conducive for him to do so. He knew before he even got out of college that this operation would be his, and it wasn't because we were just handing it to him; it was because he deserved it. We were inviting him back, and from the get-go, we allowed him to make management decisions. You have to make it conducive for these young people. What young person wants to come back from college when mom and dad are struggling all the time and don't know where the next dollar is going to come from? That just doesn't make sense. We have to change our mindset and make it conducive for the next generation.

Eliminating *the* Need *for* Synthetic Fertilizers

DR. KRIS NICHOLS

Gabe was putting the tools into place that were stimulating the biological activity, but he still wasn't fully optimizing the efficiencies of the organisms that were in the soil—especially the mycorrhizal fungi. That was because he was outsourcing the job of the mycorrhizal fungi.

Soluble fertilizer is going to do the most damage as far as outsourcing the jobs of the microbial community. So, as much as possible, if you can, avoid those soluble inorganic types of fertilizers.

Imagine that you fulfilled the nutrient requirements for the plant early on in the growing season, and the plant didn't require the presence of the mycorrhizal fungi, so you didn't get the growth of this hyphal network. Then later in the growing season, this hyphal network starts to develop, but now, it isn't very extensive because it hasn't been growing for very long. Then the plant has a nutrient demand, and the soil is very dry. Now you need to utilize more water in order to be able to try and get those nutrients flowing.

It's become a very efficient process: phosphate-solubilizing bacteria get carbon from the plant, and the fungus shares some of the carbon that it got from the plant. That's how it pays the bacteria for the job the bacteria did.

I believe that we have the capacity to feed at least fourteen billion people with nutrient-dense food. This is way beyond our current population. At the same time, we can have a high quality of life and maintain and enhance our ecosystem services.

MODERN CONVENTIONAL agriculture is addicted to synthetic fertilizers. Just how much of a difference would it make if a farmer were able to stop using them? Is that even possible?

Dr. Kris Nichols thinks it is. Kris is a soil microbiologist who has spent several decades studying arbuscular mycorrhizae, glomalin, and many other aspects of soil health. Originally from a farm in Minnesota, she earned her PhD from the University of Maryland and has worked at the USDA's Agricultural Research Service in Maryland and in North Dakota. From 2014 to 2018 she served as the chief scientist of the Rodale Institute. She currently manages several different soil health consulting ventures.

Kris is passionate about developing soil fungal populations and climate-resilient crops and the negative impact of synthetic fertilizer applications—particularly phosphorus and nitrogen applications—on soil biology. This applies especially to soil fungi and has implications for developing truly regenerative agricultural ecosystems.

JOHN: Kris, you've done a lot of work looking at soil biology and its impact on plant performance and on growing climate-resilient crops. What are some of the highlights of the research that you have worked on over the years?

KRIS: I started at the University of Minnesota as an undergrad and was able to do some laboratory research on arbuscular mycorrhizal fungi. These are organisms that have relationships with over 90 percent of plants, and most crop plants are associated with mycorrhizal fungi.

These fungi help the plants get nutrients and water from the soil in exchange for photosynthetically derived carbon. This is a very old, very important relationship that plants and fungi have had. It was fascinating to see how interconnected these two organisms could be and how much life depended on having these organisms work together. That led me to want to continue to study these fungi and plant relationships for my graduate work.

I grew up on a farm and was involved in agriculture in various ways. This was in the eighties, a time that was very difficult for farming and agriculture. Like many young people, I didn't want to do what my parents were doing and wanted to run away from the farm.

Although I was fascinated by these fungi, I didn't want to get into farming. But life becomes very funny, and I ended up falling back into agriculture when I got a position working for the USDA Agriculture Research Service in North Dakota. I was able to work with farmers, like Gabe Brown, who were doing really fascinating, innovative things—taking some of the science that I had been learning about and researching in the laboratory and actually applying it in the real world.

I was seeing that the things that I had been discovering and thinking about in the laboratory actually had connections in the real world. The things that I was questioning and couldn't fully understand—I could start to see those questions play out in what these farmers were doing.

JOHN: When I interviewed Gabe, he said that he's not using any synthetic fertilizers on his farm, and this is due to a comment that you made when you visited his farm. You said that as long as he continues to use synthetic fertilizers, he can never expect to regenerate soil and plants to their full potential.

I'm really fascinated by that comment. Could you describe why you believe that to be true?

KRIS: That goes back to the research that I was doing in the laboratory. We would see this very strong connection between the mycorrhizal fungi and the organisms that were in the soil and the potential that these organisms have for being able to resolve the nutrient issues that plants have.

In the laboratory, we could see that they could resolve those issues. We tried to get farmers to apply these methods in the field, but their systems were still very much dependent on synthetic fertilizers. I couldn't fully understand why they thought they needed those synthetic fertilizers when I was seeing in a laboratory that they didn't.

When I looked at Gabe's system and the way that he was applying things and the way that he was stimulating the biological activity, I started to think that he was putting the tools in place that were important for biological activity—reducing the amount of tillage that he was doing. In fact, he had gone to a no-till system—eliminating soil disturbance, improving the crop rotation, utilizing cover crops, making sure that he had plants growing as much as possible.

He was putting the tools into place that were stimulating the biological activity, but he still wasn't fully optimizing the efficiencies of the organisms that were in the soil—especially the mycorrhizal fungi. That was because he was outsourcing the job of the mycorrhizal fungi. They will provide the nutrients to the plants if the plants are required to give that photosynthetically derived carbon to the mycorrhizal fungi—and the consortium of organisms that work with the mycorrhizal fungi—in order to get those nutrients.

If you don't start that early enough in the process, you're not going to be able to optimize that functionality. That was the missing link for Gabe—that he needed to really optimize that functionality in order to be able to maximize those efficiencies.

JOHN: We have observed that many vegetable farmers will apply a soluble phosphorus fertilizer at planting. It's standard practice, as I'm sure it is as well on many broadacre crops. With the vegetable crops, we know that when we apply soluble phosphorus fertilizers, we don't see the mycorrhizal colonization happening.

I think what you're describing—if I'm understanding it correctly—is that when you apply the soluble phosphorus fertilizers, you essentially remove the need for the plant to have a symbiotic relationship with mycorrhizae because they can already obtain a generous supply of phosphorus.

KRIS: Exactly. Again, it becomes very critically important to work a lot with the timing. If we are going to be adding amendments, we need to time them appropriately with the physiological needs of the plant so that we make sure that the plant is getting its needs realized when the plant has those needs.

When it's very young, the plant will essentially set what its maximum can be. But then it has a period of time in which it can't fully utilize the amount of fertilizer that we apply in many of our systems because the seedling is very small. Later on, when it goes into the reproductive phase, it has a very large nutrient demand in order to fulfill that maximum potential—the grain-fill potential or its maximum yield potential.

We want to figure out how to work with the timing of the physiological needs of the plant and make sure that we've timed it in such a way that we are not outsourcing the job of things like the mycorrhizal fungi too early in the growing season. We need to keep the mycorrhizal fungi and the consortia of the organisms available to colonize the roots.

We want to make sure the plant has enough nutrients starting off so that the seed can germinate and grow in our fields—so that young seedling can realize its optimum potential and not be so overfertilized. If, as a plant, you don't need to have mycorrhizal fungi or other microbes feed you—because you have all your nutrients already supplied to you—why would you pay for something you don't need?

Everything is supply and demand. Nothing is free. We're taking a systems approach, and nothing is free in this system. We want to create a system in which the plant relies on those organisms so that it allows those organisms—the mycorrhizal fungi—to colonize the roots early on, paying them a little bit of photosynthetic carbon so that they'll colonize those roots and start supplying the nutrients.

Also, in addition to supplying nutrients and water—phosphorus as well as a number of other nutrients, including many micronutrients—the mycorrhizal fungi can help with disease prevention—especially root-borne pathogen prevention—if they colonize the roots early enough in the growing season. It can be a really important crop protection system.

They're able to optimize their efficiencies because they've been growing with the plant and building the consortium activity throughout the growing season. They're helping cycle nutrients in the rhizosphere environment—providing the right nutrients at the right times when the plant has those demands.

When the plant has the greatest demand—later on during the reproductive phase—that's often the most stressful time for the plant. That's also the hardest time for nutrient flow in the soil environment because in the places we grow our plants, this is frequently also the time our soils are the driest. Much of our nutrient movement, as well as a lot of biological activity, is dependent on soil moisture and water.

You need to make sure that you've got a moist soil environment, but we don't have a lot of precipitation events at this point in time. This is midsummer. What we want is to have these organisms in the rhizosphere. You have to have nutrients or organisms moving around in large aqueous environments. Because it's happening right around the roots, you don't have to have a lot of water in that environment. Now you've got a great level of efficiency.

JOHN: You spoke a bit about the negative impact that phosphorus can have on mycorrhizal colonization and—possibly, I would expect—also other microbes. A question many growers have asked me is what other fertilizers might have a potentially damaging effect? Is it all soluble fertilizers? Which should we particularly try to avoid?

KRIS: For the mycorrhizal fungi, phosphorus has been a key soluble fertilizer. Soluble fertilizer is going to do the most damage as far as outsourcing the jobs of the microbial community. So, as much as possible, if you can, avoid those soluble inorganic types of fertilizers.

The organic fertilizers are not going to have that same level of issue because, for them to work, they have to be broken down by the microbial communities. Those nutrients are being released by the microbial communities and then processed and cycled and becoming available.

In addition to that, we are finding that some of the proteins and the amino acids are possibly being taken up. A nitrogen that is in a protein or an amino acid form can be taken up in some smaller quantities directly by the plant's roots. We're not quite sure if that happens, but we have seen very good evidence that the fungi can act as a mediator to help take up those proteins and amino acids, translocating them inside the plant roots themselves.

JOHN: The piece that you just mentioned—plant absorption, root absorption of amino acids and proteins—that's a really big deal. Farms that have a generous supply of organic matter that we've worked with have grown greater than three hundred bushels per acre of corn without a single pound of added nitrogen. Obviously, we had the nitrification cycle working in the soil profile, and I'm certain that even on those soils, there was ammonium and nitrate present even though none had been applied.

But in the testing that we have done, it also appears that plants are absorbing amino acids and peptides, and those seem to contribute a great deal of energy to plant growth—much more so, proportionately, than nitrate or ammonium. I think that's an opportunity to grow higher-yielding crops that we have not understood very well.

KRIS: I would agree with that. There are many things that we discuss as scientists that we don't always think about taking out into the field. We don't always make that connection to what's happening in the field and how some of these more complex activities may be occurring. We just accept what we think as principles or as dogma or as "This is just the way it is—this is how things function."

We don't go out there and ask new questions and challenge ourselves often enough. We need to ask whether the plant roots have ways of being able to absorb some of these more complex molecules. We always thought that they absorbed very simple inorganic ions. If they're able to absorb some of these more complex molecules—or if that's mediated by microbes, especially by mycorrhizal fungi—we have to recognize that this is something that we're starting to see actually happening.

For me, what becomes critically important about this is that it also gets us away from the continuous argument for the use of synthetic fertilizers. It's always been that you have to use synthetic fertilizers—you have to add in. If you take something off, you have to put new things in, and they have to be in this inorganic form because that's the only way plants can take them up. You have to do this; otherwise, you're going to mine your soils.

I said this to Gabe—as far as not adding synthetic fertilizers. There was never really a question in my mind or in Gabe's as to whether he was going to mine his soils.

JOHN: When you look at the geological profile of many of our agricultural soils, most have a virtually inexhaustible supply of most of the macronutrients—particularly phosphorus and potassium. Gabe is a great example—he has built his biology and soil profile. The level of available nutrients in his soil has continued to increase to the point where, today, they're four times higher than on surrounding farms and multiple times higher than it was when he started.

KRIS: Exactly. Those are the things that we have to be saying to ourselves—that we're not going to use synthetic fertilizers. But because we've always focused on that inorganic component of things, when you do something like what Gabe is doing, if you're not adding inorganic fertilizers, people say you're going to mine your soil—your yields are going to decline, and you're going to lose something.

I don't like arguing with people. I like to change the direction and the thought pattern into "This is what's going on. If I can show you this other evidence of what's going on, there isn't really an argument to be had. We're not going to argue about this."

I'll talk to people about organic matter. There's been a long discussion of the contention that we can't change organic matter percentages in our lifetime. Many soil scientists that I have a tremendous amount of respect for still believe this to be the case.

I'll have discussions with them, and I'll say, "I understand why you have that point of view. This paradigm that your point of view is based on is this idea that the old humic material that we've always focused on—that is the majority of organic matter—can take hundreds of years to form. That's a highly decomposed product, and that's the majority of the organic matter. If that's what most of the organic matter is—if it's a highly decomposed product—that can't have the binding capacity, the chemistry, and the functionality that you're attributing to organic matter."

JOHN: To a point that you made earlier, agriculture—as well as some other areas of science—seems to maintain an abundance of dogma. Today we have a very substantial amount of evidence—large-scale, field-scale evidence—that we can, in fact, build organic matter quite rapidly. In light of that very practical evidence, we need to reconsider some of the theories that have existed before and see where we need to reevaluate.

I'd like to go back to the comments that you made earlier regarding water and water-use efficiency. It's my understanding that mycorrhizal fungi, and possibly other fungi as well, can access water in the soil profile that plant roots might not be able to access and can make it available to them. They also perhaps contribute to crop resilience in other ways. Can you describe that for us?

KRIS: You're exactly right. There are two different processes that are going on. One is the diameter of the fungal hyphae. Because they are much smaller in diameter than the roots, they can actually colonize new zones in the soil environment and access water that plant roots normally can't.

Researchers in more arid environments have been able to see this. We can see this in the natural world. Fungi are a part of allowing plants to grow on rocks. They break down rock material. They penetrate through that very dense environment and access water in very solid, rocky types of environments.

This is why you see lichens growing on rocks and allowing trees to be able to grow basically on the sides of mountains. Fungi are allowing the roots to be able to grow there and to access and break through that material. It's that diameter of the fungal hyphae penetrating through that dense material.

The other way this happens is through the overall water-use efficiency of fungi. We don't fully understand soil and soil biology and plants and how plants grow. One of the most basic things about plants is how they use water, yet we have no idea how much water a plant actually needs.

We water plants all the time. We think plants need water. We think we know how much we need to be watering plants. We've done irrigation studies. But so much of that is more dependent upon the soil in which we're growing the plants than on how much water the plant actually needs.

There are water needs on a cellular level. How much water is in the cytoplasm of plants—or in the xylem and the phloem plants utilize to move minerals? How much water is then actually needed in the media—in the rhizosphere surrounding the plants—in order for nutrients to get from that media into the plant roots themselves?

This is where it's critically important to understand and recognize that if you can get the relationship of the mycorrhizal fungi occurring early enough in the plant growth, it's going to allow you to create this network of fungal hyphae, which can be hundreds of miles long but which may only be a few feet around the plant roots themselves.

They fill that environment so that—wherever there may be an atom of nitrogen or a molecule of nitrate or an amino acid or a potassium ion or a copper ion or whatever the plant needs—there is going to be a piece of hyphae nearby. It makes sure that an ion only has to move a very short distance in water to reach the hyphae to be absorbed. And once it gets into the hyphae, it's all moving through water—water that's within the cells of the fungi. Then, once it reaches the plant roots, it's within the cells of the plant.

Now you're efficiently using water. You only used a small amount of water in the soil environment. The more you can have this type of architecture designed, the more efficiently you're utilizing water in that environment.

Imagine that you fulfilled the nutrient requirements for the plant early on in the growing season, and the plant didn't require the presence of the mycorrhizal fungi, so you didn't get the growth of this hyphal network. Then later in the growing season, this hyphal network starts to develop, but now, it isn't very extensive because it hasn't been growing for very long. Then the plant has a nutrient demand, and the soil is very dry. Now you need to utilize more water in order to be able to try and get those nutrients flowing.

JOHN: Now you're dependent on water that is in the bulk soil environment—that the plant roots can access—not on that which the mycorrhizae can access.

You mentioned another very key piece: that when the mycorrhizal network gets established at the beginning—early in the plant's life—you end up with a mycorrhizal network that substantially exceeds the root-system size. You're extending out past the root system—in some cases, quite a distance. They can access water that the plant roots themselves wouldn't be able to reach as well.

KRIS: Exactly. You're gaining layers of efficiency. This is what's been so critical about this relationship and why these relationships have lasted for as long as they have. They have fulfilled these roles and continued to make themselves more and more efficient.

One of the unique relationships that has been established is between mycorrhizal fungi and another organism: phosphate-solubilizing bacteria. These bacteria colonize the hyphae of mycorrhizal fungi. They essentially grow on the hyphae of the mycorrhizal fungi. This is good for them.

Bacteria are even smaller than the fungal hyphae themselves, which means that it's very hard for them to move to a new environment. So they hitch a ride on fungi so that they can get to different environments. They produce enzymes that release phosphate from minerals. Phosphate likes to be bound in soil to either iron or aluminum, or to calcium. They produce an enzyme that releases that phosphate.

The phosphate will be solubilized. It will stay soluble for a brief period of time and can become a compound again. But because it's solubilized near the fungal hyphae, the fungi can readily absorb it and then transport that through the fungal hyphae to the roots of the plant.

It's become a very efficient process: phosphate-solubilizing bacteria get carbon from the plant, and the fungus shares some of the carbon that it got from the plant. That's how it pays the bacteria for the job the bacteria did.

Again, because this happens right outside of the hyphal wall, it takes very little water. The bacteria may only be a couple of micrometers from the plant, but this is like being several miles away in human terms. If this process were on our scale, the fungus and the phosphate would be solubilized in the neighboring town; now you need to figure out a way to get a column of water to flow from the neighboring town to your farm to bring the phosphate. Instead, you now have the bacteria at the end of your driveway, fixing the phosphate for you, so that all you have to do is get the column of water to go to the end of your driveway and bring it to your house.

JOHN: This is fascinating because what you're describing—if I'm understanding you correctly—is that mycorrhizae not only have a symbiotic relationship with plants but also in the other direction, with phosphate-solubilizing bacteria. They are not only providing carbon and sugars; they're also providing water to the bacteria. Is that correct?

KRIS: They're not necessarily providing water. They could provide water if water were needed. The water comes with the carbon. The difference is between saving water for phosphate mobilization—from having to utilize the amount of water it would take to mobilize phosphate from the end of your driveway—as opposed to mobilizing phosphate from the neighboring town.

JOHN: Much less is required.

KRIS: Exactly—when we rely more on the microbial community.

What I was trying to tell Gabe was that he was starting to tap into these systems. We know that there's a similar connection that mycorrhizal fungi make when they cocolonize. We'll have roots of a legume and a nonlegume plant that will be colonized by the same fungus. A strand of hyphae can colonize a legume plant and nonlegume plant and can transfer nitrogen from the legume plant to the nonlegume plant.

Again, there's no water loss, because all of that nitrogen fixation is happening in the roots of the legume plant. It's all flowing through the water that's in the cells of the fungus and the plant roots. It doesn't waste water. For Gabe—who was growing things in a semiarid environment—if he truly wanted to make his system as efficient as possible, he needed to quit providing synthetic nutrients and optimize what the organisms in the soil and mixed cultures of plants could provide.

We know that plants like flax and buckwheat are also important in being able to help to move nutrients between other plants in the whole system. Flax is really important. It's not a legume, but we have seen, with labeling studies, some micronutrients move between flax plants and other plants through mycorrhizal fungi as there have been demands.

JOHN: It's intriguing that we have plant-to-plant transfer of nutrients and energy sources.

On all the various project and research that you've done, what has been something that really surprised you?

KRIS: The thing that really surprised me happened indirectly. I knew it, but it didn't hit me as strongly until a few years back. It was actually something that happened on a personal level.

I ended up having thyroid cancer. Your thyroid controls your metabolism—your body's ability to process carbon. I had to get my thyroid removed, and then I needed to wait for a while and not allow my body to have the synthetic hormone that they normally give you so that your body knows how to metabolize carbon. If there were any remaining cancer cells in my body after they removed my thyroid, they wanted those cancer cells to replicate so that when they did, the radiation treatment could kill them all.

What happened was my body started not being able to function very well. I couldn't think very well. It made me think about how the whole entire agroecosystem functions. Going back to carbon and carbon flow through all of these organisms—this whole pay-as-you-go type of system—this carbon exchange was so essential because without carbon, nothing works. If you can't get carbon flowing from one organism to another organism, it doesn't function.

Going back to the work that I had been doing with farmers like Gabe—he was doing the no-till, he was doing the crop rotation, he was doing the cover crops and those types of things, but he was missing the fact that he was cutting off the payment of carbon to the soil organisms.

Everything from then on—discussions I would have with farmers—would be about what they were missing that was stopping the payment flowing to the right organisms in their system. Are you growing cover crops? Is your rotation complex enough? Are you doing tillage so that the system has to be in a constant rebuilding process, as opposed to an enhancement process? How is the carbon flowing and being metabolized?

JOHN: What do you believe to be true about modern agriculture that many other people don't believe to be true, or that is perhaps not mainstream?

KRIS: I believe that we have the capacity to feed at least fourteen billion people with nutrient-dense food. This is way beyond our current population. At the same time, we can have a high quality of life and maintain and enhance our ecosystem services.

JOHN: I absolutely agree. I would say, in fact, that we already produce enough food to probably feed twelve to fourteen billion people. But we have a distribution problem—not a production problem. There is so much untapped potential. Every crop that we are growing has untapped genetic potential. With the genetics that we already have today on many crops, there is the potential to increase yields anywhere from 30 to upward of 100 percent.

I completely agree that we have the capacity to do so much more. We already have the knowledge. We have the information. There's still a lot that we don't know and a lot that we need to learn, but we already know enough to make a really substantial difference.

KRIS: I totally agree with you as far as how much we produce. I think the biggest thing that we're missing now is the nutritive quality of what we're producing. Many of our important amino acids—antioxidants, what we've always referred to as polyphenolics and phytonutrients—are actually more tied to soil biology and soil microbiology. That's what we need to be tapping back into.

JOHN: I'd love to unwrap that a bit more because this is something that we've observed, and I haven't seen it being explored very well in the literature. We have observed that when we have really aggressive biology in the rhizosphere profile that there is a much greater production of what are generally referred to as plant secondary metabolites—all the terpenoids, the phytoalexins, and all these bioflavonoids and aromatic compounds that plants produce.

Based on the work that you have done, what is the impact of biology on producing higher levels of these compounds in plants?

KRIS: Well, there are two things that are going on. One is that there are many compounds that are actually coming directly from soil microbes—especially many fungi, which produce many of these compounds. Then those compounds are finding their way into the food chain through plants. The plants are taking in many of these compounds that are actually originally produced by microorganisms.

JOHN: That would be fascinating because that's another example of plants absorbing complete compounds rather than ions.

KRIS: Right. We're not quite sure how this happens. Some of this we do know is mediated by fungi, especially root-colonizing fungi or the arbuscular mycorrhizal fungi or the ectomycorrhizal fungi.

There's a compound that we started working with when I was still at the Rodale Institute called "ergothioneine" that has been found not just in plants but also in animal tissue—in chicken livers, in parts of animals that find their way into what we're eating. This is because those animals are eating plants that have high concentrations of these fungi growing in the soil.

We're trying to understand how we can enhance those relationships—what management techniques impact those relationships. These are secondary compounds that are being formed. It goes back to this whole idea of carbon and where carbon flows and how carbon moves in the system.

Especially on a microbial level, and especially for fungi, when your primary job is to satisfy the nutrient needs of the plant, the first thing you're going to do is make sure that you can provide the maximum amount of nutrients that the plant needs. I think that's why mycorrhizal fungi are tied to something like phosphorus—because phosphorus is so unavailable in the soil, and it's a nutrient that plants need in abundance. It was something that we could see a pretty direct relationship to.

When it came to some of the micronutrients—and to water and to some of the other things that mycorrhizal fungi also provide to the plant—it took a little bit more time and a little bit more research to take a look at. It's not that those things aren't as important as phosphorus; it's just that we started looking at phosphorus first because our focus, when it comes to plant nutrition, has always been on the big three: nitrogen, potassium, and phosphorus. Phosphorus was always a tough one in the soil because it's immobile. That became a big key.

We did these studies where we manipulated the concentration of phosphorus and looked at how that affected colonization. We saw how it affected plant growth. But we're starting to see now that other things are actually happening as well.

When the fungus is reliant upon satisfying nutrient needs, the first thing the fungus is going to put energy into is creating a better hyphal network—extending that network as far as it needs to in order to make sure that it gets its nutrient demands satisfied so that it can get as much carbon as it can. Then, if it has the network established well, it starts to do some of the secondary activity—enhancing that network a little bit more, making sure that part of that network has some rigidity to it.

Some of the hyphae will invest in making secondary compounds as part of the hyphal wall structure so that it doesn't fall apart as easily. Most of the hyphae, when first created, are very ephemeral. They don't last very long. They don't put a lot of energy into it.

If it's going to stick around a little bit longer—if there isn't much disturbance with tillage or things like that—then it builds up a more resilient network. It starts making soil aggregates with that more resilient network. It starts investing in releasing more carbohydrates—more sugars to attract phosphate-solubilizing bacteria and consortia of other organisms—creating what are referred to as "mycorhizosphere organisms" around the fungal hyphae. These start to develop more of the soil aggregate structure.

Part of creating that more rigid hyphal network is creating some of these secondary compounds. The formation on the part of the arbuscular mycorrhizal fungi—some of the initial work that I had done my PhD and master's work on—was looking at glomalin. Glomalin is a glycoprotein produced by arbuscular mycorrhizal fungi.

We found that glomalin wasn't always immediately produced. Systems that were no-till—or undisturbed systems that used perennial crops or cover crops—produced more glomalin. It's one of those secondary compounds. It requires more carbon in order to be produced.

That makes sense. It's a large, complex biomolecule. If the first thing you need to do is make sure that you do your primary job—getting nutrients—then you make a good network in order to get the nutrients. If you make a good network, then you now have the energy reserves to make secondary compounds. And glomalin, from the perspective of the fungus, helps with some of the hyphal wall rigidity. It potentially allows the hyphae to penetrate air-water interface zones.

That's something that we find with those secondary compounds. What they do for hyphal wall structure is also what helps us classify them as antioxidants and polyphenolics and some of the phytochemicals that are important for human cell wall integrity and some of our own health benefits.

JOHN: Kris, we've spoken of a number of different topics: mycorrhizal fungi, soil microbiome, and many different pieces that are tied together. I'm certain that growers will want to explore some of these topics more deeply. Are there any resources that you recommend frequently to growers?

KRIS: I encourage growers to go out there and really explore a number of different types of resources. I do a lot of reading through scientific literature, so you can plug your way through that. But there are a few authors who have done a good job of summarizing some things, and they have some good references that will point you to some scientific literature.

David Montgomery is a very good author and has some good reference lists. And Judith Schwartz, who wrote a book called *Cows Save the Planet*, has also done a book called *Water in Plain Sight*. It's a really interesting book about water relationships and some of the other issues that we've touched on—again, providing some reference materials. And Kristin Ohlson's *The Soil Will Save Us* is also good.

One of the books that changed my life was *Biogeochemistry* by William Schlesinger. It can be a difficult book to slug your way through, but it opens up some interesting topics that interface not just biology but biology and chemistry and geology and physics. I was just telling a colleague of mine the other day that I consider myself to be a physical biogeochemist because—being a soil scientist—that's kind of what you have to do: look at all of these interfaces.

That's what I would encourage growers to do. I've always thought of farmers, ranchers, and growers as Renaissance people—people who function at the interfaces of so many different occupations. It's hard to point to one particular topic, but explore a number of different things and a number of different resources and how they all interface and work together.

Ecosystem Diversity Prevents Insect Pressure

DR. JONATHAN LUNDGREN

The question that I always ask farmers who come visit is, "How often can you lose a crop and make more money than the neighbors do?" And the answer is, "Every single time when you're farming regeneratively."

Entomologists always focus on a couple of key natural enemies; we'd never really looked at the actual community in all of its glory. In looking at the data, the patterns became so clear: that when we had more insect species within a cornfield, there just weren't pests. We did not have "pest" problems anymore.

Look at how we produce our crops right now—if diversity is what we're after, then our production practices are just absolutely perfect for creating pest outbreaks.

If you have a pest outbreak in your field, that means that you're doing something wrong. That's a very, very different philosophy than you get from just about every extension agent or from the current scientific infrastructure. This infrastructure is bent on making a broken system work.

Some of the best naturalists I've ever met are farmers and beekeepers—observers of the natural world.

WORDS AND PHRASES ARE POWERFUL. Think about the phrase "food allergies." Why have we come to accept as normal that people should be allergic to food? Dr. Jonathan Lundgren brings a similar phrase to mind: "food safety." Dr. Lundgren brings an interesting perspective as a professional evaluator of pesticides and genetically modified organisms in agricultural ecosystems.

Dr. Lundgren earned his PhD from the University of Illinois in 2004 and served with the USDA's Economic Research Service for eleven years. He is the author of hundreds of peer-reviewed papers, as well as the book *Relationships of Natural Enemies and Non-Prey Foods*. He is the director of the Ecdysis Foundation, which seeks to "support the reimagining of agriculture along ecological principles with independent, farmer-driven research." He also serves as the CEO of Blue Dasher Farm in Estelline, South Dakota.

Dr. Lundgren asserts that we have the potential and the capacity to develop agricultural ecosystems that are incredibly resilient and that we can produce crops that are completely resistant to insect pests.

JOHN: Jonathan, you've done some very interesting work looking at agroecology and the ecological implications of pest management strategies and some of the tools that we use today. What are some of the memorable moments that led you to the spot where you are today?

JONATHAN: I was trained as an entomologist. A lot of times, scientists are trained to focus in their study area and within their particular group to specialize, and I was no exception. I am a predator ecologist, and I really liked lady beetles—ladybugs. But the deeper I dug into these beetles—to that organismal level—I started to ask questions that couldn't be answered at an organismal level. That forced me to start to expand my thinking about the research that I was focusing on, including more of a community approach. The more I focused on insect communities, the more I realized that in order to answer some of the questions that needed to be addressed, it required that I not just think about entomology anymore and start learning more about the agroecosystem in general.

I still study insects a lot, but I realize now that they're good learning tools for telling us about the greater system rather than being an entity in and of themselves.

JOHN: When you speak of insects as being an indicator of a greater system, what are they telling you? What have you observed that surprised you?

JONATHAN: One major change in my thinking is realizing that I can't be focusing on one particular group of insects. We all want a dipstick to tell us whether our practices of soil health are working, right? What I came to understand—the more we researched this topic—is that insect communities were an entity that we needed to be focusing on rather than individual indicators. Insect communities were so dynamic and they changed pretty dramatically, but at the same time, all of those changes were really representing key niches. Diversification of that community allows the community to function as we change the habitat or the farmland. There's always somebody sitting in each piece of the puzzle, and that's where biodiversity really comes into play. That was what was so surprising—just how important biodiversity within our agroecosystems really is. We need to like it in all of its forms, don't we?

JOHN: Yes, we do. What did you observe that really drove this point home for you?

JONATHAN: We started a bioinventory of cornfields. I began my career as a scientist with USDA. I swore that I wasn't going to work on corn because a lot of entomologists worked on corn at our laboratory. Corn has been the system that really opened my eyes to the importance of diversity, though.

We landed a small grant to study insect communities in corn and how landscapes affect those communities and the resilience of those communities to invasive species in cornfields. We led this whole team—part of it was entomologists, and the other half was the economists from all over the world. My colleague in the project was an economist. We surveyed fifty-three cornfields all over Eastern South Dakota, and in every cornfield, we applied the same methods, and we looked at all of the insect species that were living on corn plants. We collected them up and then we brought them back to the lab. It was a huge, monumental effort—this was a lot of work. We ended up identifying, by the end, probably hundreds of thousands of insect specimens, down to the species level.

The more I dug into it, the more it opened my eyes to the fact that this had really never been done before. Entomologists always focus on a couple of key natural enemies; we'd never really looked at the actual community in all of its glory. In looking at the data, the patterns became so clear: when we had more insect species within a cornfield, there just weren't pests. We did not have "pest" problems anymore.

At the same time, there were a couple of other papers that were starting to come out that were suggesting that this lack of diversity was really important in creating pest problems. Look at how we produce our crops right now—if diversity is what we're after, then our production practices are just absolutely perfect for creating pest outbreaks.

JOHN: I can completely relate with your desire to not work with corn—it's a dominant crop, it's well studied to some degree, and it's a relatively simplistic crop compared to growing, say, tomatoes or peaches or nectarines. In the state of Iowa, any crop that is not corn or soybeans is considered a specialty crop. Alfalfa hay is a specialty crop. Is this not amazing?

JONATHAN: Pretty amazing. I can justify working on it in my own mind by just saying, "Well, if we can make changes to corn, think of the footprint geographically that we would have in terms of changing landscapes." On our own farm, I don't grow corn, and I don't advise people to grow corn either.

JOHN: The reality is that for many growers today, it makes a very good economic sense to grow crops other than corn. But that's another conversation.

You mentioned that when you observed increased insect diversity, you had no pest pressure. What do you believe the mechanisms are for producing that shift?

JONATHAN: We've known that putting up more plant species into a habitat tends to result in fewer pests. There are two general hypotheses for what drives this: the natural enemies hypothesis and the resource concentration theory. The natural enemies hypothesis says that as you increase plants, you increase the number of predators of pest species. That predation ends up bringing a downward pressure on the pest populations. The resource concentration theory says that by including more plant diversity into a system, you dilute your focal crop within the vegetation community. This might make the focal plants more difficult for the pests to find or change them in some way. I think both of these ideas are probably at play within the study that showed that insect diversity was really important, but I think it's so much more complicated than that.

Scientists are just like everybody else—we like to categorize things, we like to control them, and we like to manipulate them and feel like we have some intimate knowledge of what's going on. But the reality is that within ecological systems—and I've been working on them for a long time—the complexity becomes so staggering that I don't think we can understand all of the mechanisms. That's a humbling experience—for a scientist to have to admit that at some level we just have to get out of the way and observe the systems, rather than going out to our research farms and picking out every mechanism that we can think of as the "experts."

What I ended up doing that was really important in my career—and in coming to these conclusions—is that I started to go out to the farmers that were leading this regenerative agriculture movement and looked at what they were doing. I wasn't interested in controlling their farm. I just wanted to observe. That's the way science used to happen, isn't it? Scientists would go out, and they would observe the natural world, and they would use that as the basis of developing hypotheses that they then could test.

But nowadays, science has totally shifted. Today, the scientific expert who's got a PhD and all the pieces of paper that say that they're really bright, they sit down behind their computer, and they write the sexiest grant proposal that they possibly can. There's a total disconnect, more often than not, between the science that ends up happening and the real world that farmers are ultimately living in. It's been a real challenge for us to innovate our food production system using science.

JOHN: It's a challenge for some growers as well. There are certainly some growers who are exceptional at being observant and seeing what is happening. Many farmers are very connected to their farm and what's happening around them in the natural environment. There is an incredible complexity. It's beyond the capacity of most of us—or perhaps any of us—to comprehend all the interactions that are happening inside of a plant. With all of these various organism communications back and forth—by whatever mechanism and by whatever means—there is beauty within all of that complexity. We can take comfort in the fact that it was designed to work. It's been developed to work. It works extremely harmoniously. And to the point that you made earlier, in many cases, perhaps we simply need to get out of the way and allow it to function as it was intended to.

JONATHAN: To me, it's humbling, and it's also so exciting because when we start appreciating the beauty and that sense of wonder at the natural world—what an invigorating thing!

JOHN: It is, indeed.

Earlier in our conversation, you mentioned the ecological implications of pest management and some of the tools that we use today. I'd love to hear a bit more about that. What are the impacts on our agriculture ecosystems of the pest management tools that we're so widely using today, and what are possible alternatives?

JONATHAN: My research focuses on two general areas. One is risk assessment of pesticides and genetically modified crops and, basically, farm management practices in general. The other half is working on developing sustainable systems. I've got quite a bit of experience in understanding the ecological risk assessment in the framework that we use for trying to determine whether a particular agrochemical or something is safe or not safe. I've advised the US EPA. I've been on advisory panels for the European version of the EPA, the Brazilian government, the United Nations, and their conference on biodiversity. After twenty years of studying this, I can honestly say that when you look at the side of a (pesticide) jug, and it says that it's safe, that is meaningless.

You can go out, and you can buy something off of the shelf, and it's labeled and it's regulated. Nobody is watching. Nobody is watching! Risk assessment is incredibly hard. We can get closer to the mark with scientific approaches to risk assessment, but again, when you understand the complexity of the natural world...!

Right now, just for pesticides, there are a couple hundred different pesticides whose active ingredient is registered with the EPA. That's where most of the safety assessments are focused—on those active ingredients. But whenever you add an adjuvant—maybe it's a sticker or a spreader or a defoaming agent or whatever—it changes the toxicology of that active ingredient such that the risk assessment that you're actually performing on an active ingredient really does not hold much weight anymore. Think about all of the formulated products—there are twenty thousand formulated pesticides in the

US, and each of those pesticides would require an independent ecological risk assessment—each of them. Where's the budgeting for that?

JOHN: Farmers will, in many cases, add additional adjuvants or spreader-stickers to the spray tank, in addition to whatever is in the jug.

JONATHAN: That's right. That changes everything. I'm not saying that we shouldn't have agrochemicals, be they herbicides, fungicides, or insecticides. I understand the utility of these things, but I think we need to have a whole lot more respect than we do right now. The counties—my county in particular—should not be using a fire hose to spray the weeds on the road margins to kill weeds that may or may not be there just to burn up product. Come on! When farmers buy a bag of seed, and the seed is an unnatural color—it's either purple or blue or pink—those are neurotoxins that they're putting onto their field. Very little of that is actually getting into the plant. Most of it is getting into their soil.

What causes pests is a reduction in biodiversity. When you buy a jug, you are counteracting nature's ability to fight that pest. You are causing pest problems. Then you have to buy another jug in order to control those pests. It's this wicked treadmill that just keeps spiraling.

JOHN: When you say that nobody is watching, what are the ecological implications of that? What I'm hearing you say is that we don't even understand the implications that these materials might have on soil health and ecosystem health.

JONATHAN: We don't have the first inkling of what the implications of these things are, be it glyphosate, neonicotinoids, propiconazole, or the adjuvants, which are sometimes more toxic than the active ingredient in terms of their ecological effects. We're not just talking about the soil, right? We're talking about human health problems. Farmers have among the highest suicide rates of any career at this point in the US.[4] We know that pesticide use is linked

4 Matt Perdue, "A Deeper Look at the CDC Findings on Farm Suicides," National Farmers Union blog, November 27, 2018, https://nfu.org/2018/11/27/cdc-study-clarifies-data-on-farm-stress/https://nfu.org/2018/11/27/cdc-study-clarifies-data-on-farm-stress/

with depression. The science has been done on that. I mean, are we killing ourselves?

JOHN: Many of these agrochemicals, in my understanding, are also exceptionally potent endocrine disruptors.

JONATHAN: And estrogen stimulants, and the effect of this is out in the natural world in many ways.

JOHN: Endocrine disruption shapes who we are and our personality and our thought patterns in ways that we aren't fully aware of. But when you think about the cancer profile—cancers of various organs and the endocrine system, or the reproductive system, and the way that has shifted, not just among farmers but in the general population—it leads me to wonder whether all of these cancers that are connected to the endocrine system and the reproductive system, which were a relatively small proportion of cancers a few decades ago, are related to our agrochemical use. Sixty years ago, these cancers were in the single digits; today a majority of the US population will get cancer, when taken in aggregate.

JONATHAN: Will we ever determine whether pesticides are the cause of that? I don't know. I think there's evidence that would support the carcinogenicity of various agrochemicals. I don't know that this would cause these patterns that we're starting to see within our society. I'm not sure, but boy, my eyes and ears are wide open. If on my own farm I can get away without using that stuff, then I'm going to do my darndest to do that.

Perhaps more importantly, I don't try to make it an ideological decision. If I can eliminate an input cost from my farming operation, that's just a good business decision. I tell farmers, "There's another way, and it's probably going to save you money."

JOHN: What is that other way? How do we develop agroecosystems that reduce the need for insect management products and other agrochemicals?

JONATHAN: The regenerative farm movement is doing a heck of a good job of it. There are some principles that link these systems. Fundamentally, reducing or eliminating tillage from your operation is the first step. Never leaving your soil bare is the next. Increasing plant diversity whenever possible on your field—be that crop rotation, be that field margins, be that intercropping, be that interseeding cover crops in your crop rows. Those are all great practices that work. And then, finally, integrating animals and crops together in the same place at the same time. When a system has these elements, they don't have pests.

JOHN: That's a really big statement.

JONATHAN: I think that is a resilient system. I can give you an example. We have a field that's been no-till for twelve years. I think it's been sprayed with 2,4-D to eliminate weeds, but otherwise, it's had perennial grass cover for a long time. We had some weeds out there, so I did use maybe a half to a quarter rate of Roundup—glyphosate—in order to just push down some of that vegetation. Then I planted our cash crop directly into that. It's an annual cash crop that's put into perennial grass cover.

Until I had livestock, I didn't understand how I was going to control those weeds in any other way. This spring I used high-intensity grazing on that weed cover, and then we planted directly in. Once my flock of hair sheep gets big enough, I don't think I'm going to need herbicides anymore. In this case, I'm actually seeing a problem—a pest problem—that we're supposed to react to. I look at that pest problem as an opportunity.

We had a tour of South Africans here last year, and they were like, "Gosh, you got some weeds out here."

And I said, "Where? I don't see any weeds. I see forage."

When you change your perspective on things, you've got real opportunities—when you don't negate the tools that you have in your tool belt. Someone will say, "I'm a cropper. I don't raise livestock." Well, you know what? You're eliminating a really valuable tool from your operations.

JOHN: Speaking of changing perspectives, I've observed that many growers and researchers shift their perspectives over time as they're on this journey.

What is something that you believe to be true about modern agriculture that is different from the mainstream view, as your perspective has shifted?

JONATHAN: When I got into entomology, I adhered to the pest management principles—you go out, you sample your fields, you wait until the pest reaches a certain density, and then you react. What I discovered, or was taught by farmers—regenerative farmers—is that if you design your system right, pest management should not be a reactive process.

Pest management is a proactive process. You design a system correctly, then you don't have pests anymore. If you have a pest outbreak in your field, that means that you're doing something wrong. That's a very, very different philosophy than you get from just about every extension agent or from the current scientific infrastructure. This infrastructure is bent on making a broken system work:

"Make it work."

"Figure out how to take this system and change it ever so slightly so that farmers will adopt it on a huge scale, and it'll get them by for a little while longer."

"Genetically modify the plants."

"Here's some seed coating."

"This will be easy to use and will make the system work."

Ultimately, the system always rebreaks. What I was taught very recently is that we need to stop supporting a broken system, and we need to reinvent how we're thinking about agriculture.

JOHN: It is intriguing how entrenched the current model has become, to the point where people are dogmatic and will no longer consider new ideas or new inputs, as might have been the case a number of decades ago.

You mentioned that in addition to pesticide evaluations, you've also done evaluations of GMOs. And you've mentioned seed coatings a couple of times. I'd love to hear what you've observed and learned about GMOs and how they impact ecosystems. Is there a role for genetically modified organisms in regenerative agriculture ecosystems?

JONATHAN: I'm not fundamentally against genetically modified organisms any more than many of the other tools that are put out there to support the current system. I think that GMOs, seed coatings, and pesticides in general are a symptom of a greater problem, and that problem is a simplification of our food system.

I'm against GMOs because I just don't think they're needed. We have to solve the underlying issue, which is a simplified system of monoculture—thousands of acres of a single plant species that's genetically limited and selected to produce only large seeds. Or maybe there are a few other small characteristics that we decided we wanted in that particular organism. Because we went down that road, we just predisposed ourselves for pest outbreak after pest outbreak. If that's your system—dead soil and a monoculture—then you are absolutely reliant on agrochemicals and genetically modified organisms. That's the only way to produce efficiently within that system.

But if you suddenly look at one of your neighbors who's farming regeneratively and you see a different approach—a different philosophy toward producing food—it's really empowering, right? Suddenly you keep more money on your own operation, rural communities keep more money in their own communities, we have opportunities for stacking enterprises and industries that keep our kids on the farm, and we're growing healthier food. Everybody wins in this situation. That's something that's worth fighting for and changing my entire life and career trajectory for.

JOHN: One of the pieces that is really inspiring to me is understanding that we have such tremendous untapped genetic potential in the crops that we have available today and have had available for the past forty years. Speaking with experts and scientists in the field—Don Huber and Jerry Hatfield and many others—the implication—in some cases, the explicit statement—is that we have this untapped genetic potential because applying more nutrients and pesticides can only push yields to a certain threshold, given our current agricultural ecosystems and our soil health.

There's all this discussion around the need to feed a rapidly growing world population—and there's a lot wrapped up in that, considering food waste and unequal distribution, etc.—but even leaving those issues aside, it is very inspiring to me to think that there are some crops where yields have increased by 250 to 300 percent over the last twenty years. Consider strawberries in California: twenty years ago, the average yield was four thousand flats per acre; today it's ten thousand flats per acre. The same is true of processing tomato production. There have been a number of crops that we have continued to develop and to get rapidly higher yields from, but that hasn't happened yet on our commodity crops—corn, soybeans, etc. The evidence suggests that the genetic potential is there—we can tap into it once we shift our ecosystems and begin growing those plans differently.

JONATHAN: We've got this list of recommendations that work for degraded soils—or at least that can keep a farmer with degraded soils going in a monoculture situation—but when you start farming regeneratively, a lot of the recommendations that are associated with that current system, including varietal selection and things like this, have to be thrown out the window. They just don't apply anymore. That's going to necessitate that we reinvent what we think we know—going back to first principles thinking, foundational thinking.

JOHN: Jonathan, what is a book or resource that you recommend most often to growers?

JONATHAN: I love some of the classic stuff like *Plowman's Folly* and *One-Straw Revolution*. They're always good. A lot of the stuff that I read tends to be a little bit more technical. Mark L. Winston's got some great stuff that I would definitely look into. He's brilliant. I love anything by Carl Sagan. I think he may have been an alien who fell down to the earth, the way he thinks philosophically. I like being challenged to think outside my comfort zone.

JOHN: Can you tell us a little bit more about what you are doing on your farm with livestock and how you're integrating livestock into your farming?

JONATHAN: Sure. We have forty-three acres. Some people would laugh at that size, but I honestly don't know that you can farm regeneratively—truly regeneratively—on anything more than a hundred acres unless you're hiring a lot of staff. Twenty-five of our acres are native unbroken prairie and wetlands.

We're going to leave that. We produce annual crops for seed, honey, and we have sheep. We're starting to integrate pastured pork. We have egg production, and we produce some broilers. It's a pretty integrated system—taking small pieces and stacking enterprises on to those things.

Last year, we grew three different crops: borage, phacelia, and hubam annual sweet clover, which most farmers have never heard of. But what they all have in common is that they're great cover crop species. They're great in pollinator conservation and things like this. They're all flowering species that kick out nectar and pollen like you wouldn't believe. And they're all worth money. Why would I plant corn, soybeans, or wheat if the market just isn't there for those things? I'd rather sell my crops by the pound than by the bushel.

These are a little bit higher risk crops. I had a difficult time trying to harvest the borage. We can grow it up through seed production, but it gets really fussy—it likes to shatter if you stare at it in a funny way. I didn't get any crop revenue off of that field, but I still made more money than my neighbor because I produced so much honey off of that field per acre.

The question that I always ask farmers who come visit is, "How often can you lose a crop and make more money than the neighbors do?" And the answer is, "Every single time when you're farming regeneratively." Every time—because you're stacking enterprises—your farm is a resilient business model. You're not tying your entire livelihood to a single commodity.

JOHN: I'm really surprised to hear you say that you don't think we can farm regeneratively on a scale greater than a hundred acres. Could you expand on that a little bit? Because I know that there are some growers who are both in the fruit and vegetable space—and in the product or crop space, such as Gabe Brown, for example—who are on a scale of thousands of acres and are making some very substantial progress.

JONATHAN: I think you can make some substantial progress, and I think that we can farm larger acreage. We can apply some of these principles. But when you're starting to talk about integrating livestock as a weed control measure—or maintaining perennial fruit and nut production, annual crops, a number of different livestock—the amount of maintenance, at least in my experience, that has to go into that system in order to keep it productive and keep it moving in such a way that that the ecological processes are hitting on many different aspects of the farm all the time … man, that's really difficult.

What's also really important to realize is that the goal of farming regeneratively at a conventional scale is probably misguided. I think that if you are farming in a regenerative nature and are stacking those enterprises and are having those different tools at your disposal, then you don't need to farm at such a great scale anymore. The average farm size doesn't have to be 400 to 1,000 acres. It can be closer to 100 to 250 acres, and you're still generating the revenue that's needed. Those are my thoughts, at least. All of those things are debatable.

JOHN: I do agree with you that regenerative agriculture ecosystems, in my experience—what I've observed—are always more profitable.

JONATHAN: We've got data that show that.

JOHN: They are extremely climate resilient. And there's something that I would refer to as the "heart-to-acres ratio." As we were discussing earlier, in order to be really observant and to really have a connection to what is happening on your farm, you need to spend time with your feet on the ground and your hands in the soil and on the plants, not just riding a four-wheeler or a pickup truck.

JONATHAN: Definitely.

JOHN: There is a limitation to how many acres you can do that well with. I also agree that farms must be "scale appropriate." This can mean different things in different places depending on the environment and the ecosystem, but generally, when you have a smaller scale, you have more intense production because you have more layers on a single operation. That can add a tremendous amount of value. I do also think it can be possible to do it on a larger scale, but you lose some of those benefits that we've just been talking about.

JONATHAN: Maybe putting a magic number on it isn't the most appropriate thing, but I think adopting the philosophy that you can farm smaller better is a goal that's associated with regenerative production systems.

I was looking at the neighbor across the road from me who was selling seventy-five acres, and I was thinking to myself, "I have forty-three acres right now, and I'm riding this all the time. Why would I expand that with another seventy-five acres?" Maybe I'll hit a certain threshold with this farm. I've only been here for three years. We'll see where it goes.

JOHN: I'm really intrigued by your description of your farming operation. Your primary crops—everything is revolving around honeybees, and you've mentioned the importance of beekeeping. What is your background with honeybees?

JONATHAN: I'm an entomologist, and bees are part of that, right? They're not crops, and most people don't think of them as livestock either. But I would beg to differ. Bees are just another form of livestock, aren't they? I'd never been a farmer or a beekeeper when we started this research farm. I advise beekeepers, I advise farmers and scientists, and we've done work in all of these things. But when I left the USDA to start Blue Dasher Farm and the Ecdysis Foundation, our nonprofit, it was in order to really serve. I had to experience what the beekeepers and the farmers were experiencing firsthand and tie our own livelihood to the regenerative farming operation.

I've learned a lot in the last three years, and I've made a lot of mistakes, but at the same time, it's given me such new perspective on just the diversity of skills that farmers need. It's also raised in my own consciousness the fact that some of the best naturalists I've ever met are farmers and beekeepers—observers of the natural world.

JOHN: I like that. That means a lot to me personally because before I started doing my current consulting work—starting Advancing Eco Agriculture and working in the regenerative agriculture field—my dream was to be a full-time beekeeper. I had over thirty colonies of honeybees from the age of twelve, and it's still a hobby today. I love it.

JONATHAN: There's nothing more invigorating than opening up a box of ten thousand stinging insects. I had some Africanized hive that were mean, but this year, they're really friendly. They can let you into this world that is almost unimaginable. Who could come up with honeybees? I mean, who could imagine something like that in science fiction or in Hollywood, right? There's something really special about that.

JOHN: Unless you really screw things up, it's always a very calming experience for me to be out and working on them because you have to be in that home space to really work with them well and not get things riled up.

JONATHAN: Livestock in general has taught me about myself—how to respond to social interactions in my own life. When the sheep get out of the paddock, they want to be back in that paddock, but you can't rush them, right? You can't run after them. You nudge them, right? You just have to take a deep breath and say, "All right, I'm going to go and do this again"—very slowly, very methodically. It teaches patience; it teaches you self-awareness. It's something that our society has lost.

JOHN: I think this is an important topic for us to speak a bit about because it's perhaps something that farmers might not hear very often. In the scope of your forty-three-acre farming operation, how many colonies of honeybees do you have, and what is the economic contribution of those as a proportion of your overall farming operation?

JONATHAN: We have a hundred hives on our land. We had to build comb over the first couple of years. That's something that a lot of new guys may not understand. I didn't. A lot of the honey initially goes into comb production. Right now, we're expecting to be able to pull down three thousand pounds of honey off of our farm, and then we sell it at eight dollars a pound, and I can't keep it in stock. Our farm is really designed to produce honey.

JOHN: That's a revenue of five hundred dollars an acre without even considering all the other enterprises.

JONATHAN: That's right. Plus, I'm getting meat production—the sheep and pastured pork—and then we're also taking the annual crop seed off when we can get it. But if we don't, that's not such a big deal. I've got a pretty good design there. The honey production is the majority of our revenue at this point. That's why we've designed it the way we have. I think that a lot of farmers, and maybe even consumers, don't even know what real honey tastes like. Honey that you buy at the grocery store—that's not honey.

JOHN: No, it resembles corn syrup more than it does real honey.

JONATHAN: Yes, absolutely it is, because a lot of the big box stores buy their honey from international markets, and China and India are dumping honey on North America right now, and it's not honey. It's like rice sugar. They've adulterated it to get through customs. They're just dumping it. That drives the price way, way down. Then when you do get clover honey, sometimes they produce that honey not off the nectar, because there isn't enough nectar in the environment anymore to support a lot of bee colonies. A lot of beekeepers end up feeding their hives sugar water or corn syrup water so that honey that's been created doesn't have the floral bouquet—the flavors and the smell of what real honey is. We don't feed our bees. We will feed them honey during June when there's a dearth, but there is almost nothing better than reaching into a beehive when it's still warm and that honey coming out.

JOHN: I'm totally with you on that. People who are aware and educated about this are aware that honey has different flavors, but the feedback I've received from buyers of honey that I've produced and that I've worked with is, "This honey is different. It makes my lips tingle. I feel energized from it." That's what honey is supposed to be like.

JONATHAN: That's right. And it's good for you. You know how good it is for you? It jazzes up your immune system. I've never kept hives alive all winter yet. I'm going to try something different this winter, but we can't keep them going. It's so discouraging to look at the learning disabilities the bees have adopted from the agrochemical exposure they're getting out in the environment. They're the canaries in the coal mine.

JOHN: I think you have the capacity to offer a unique perspective on this, with your background as an entomologist and looking at agroecology. Why are honeybees dying?

JONATHAN: You save the bees if you heal the soil. A number of pollinator groups have adopted this since I first started saying it, but it's absolutely true. The bees' die-off is totally a symptom of the simplification of our food system—the lack of floral resources: flowers, nectar, and pollen for the bees to eat. Nutritionally, they're starving.

Then you've got all of the pesticide exposure. Scientists have found ninety different pesticides in beehives. Bees aggregate these things into the wax and their comb. Their larvae are constantly exposed to this whole suite of different agrochemicals that we are being exposed to as well. And lo and behold, starving poisoned bees get sick.

Then enters in the varroa mite, the viruses, the bacterial and fungal diseases. It's a perfect storm. But you can sidewall all of those issues, right? "Oh, the varroa mite must be killing bees. I'm going to work on the varroa mite. I'm going to get rid of all these pests." We're at the stage where component research and component solutions aren't going to get us where we need to be. There's got to be a system-level change. The beekeepers get it.

JOHN: A couple of thoughts for you on overwintering. I started doing something pretty interesting three years ago, and I started berating myself as soon as I opened my mouth because I had 100 percent colony loss last year. Who am I to have an opinion? For the two years prior, I was actually successful at overwintering, and I learned a technique from an aeronautical engineer. He was attempting to remove the self-generated humidity from within the hive, and he developed this idea of removing the north sidewall frame and replacing it with a piece of marine board. He'd do this all the way down throughout the entire colony—to basically have a barrier layer of marine board between the frames and the north facing sidewall all the way down to the bottom board. The principal was that if there was any condensation or humidity within the colony, 100 percent of it would be in that crevice between the marine board and the sidewall of the colony. It does actually work that way. It removes all of the humidity and the condensation from other parts of the colony.

JONATHAN: That's been a big problem for us for sure. You were taking out a frame?

JOHN: Yes, he was taking out one frame, and then he cut it long enough so that it would fit all the way from the top of the colonies. He was using two deeps that would go all the way from the top, straight to the bottom board. He didn't want any splits, and that keeps the marine board strong.

JONATHAN: I'm going to try it. The other thing I'm going to try is overwintering them as nucs in a shed; I want to see whether or not I have to replace the resource frames every so often that way. I guess they're able to regulate temperature a little bit better. The problem—and again, I think this probably is an artifact of the neonicotinoid exposure—is that first frost. They don't know how to form a ball in order to conserve heat. You see major die-off. We lost half of the hives after the first frost. They were going strong, and then—boom—they just crashed. A couple beekeepers told me that the bees don't know how to form a ball anymore. They can't remember.

JOHN: I have extended family—cousins, aunts, and uncles—in the thumb of Michigan. They have started putting their honeybee colonies in plastic calf hutches over winter with incredible success. The last three or four years, they've had 90-plus percent overwintering success. You can think of them almost as a miniature greenhouse. They protect the bees from the wind, and they're warmer during the winter months. They will get cold, but they warm up pretty quickly whenever there's sun.

JONATHAN: Five or six in the hutch or something like that, you think?

JOHN: Yes. I want to say three or four.

JONATHAN: And that way the hive heats itself. It isn't going to get overwhelmingly hot either, is it?

JOHN: Right, exactly.

JONATHAN: What bothers me is that overwintering did not use to be this much of a challenge. Yeah, you've always lost a few hives over if you overwintered up north, but it seems like nobody is successful. I talked to all of the beekeepers in the region. Nobody had their hives survive this last year.

JOHN: Wow, that's bad.

JONATHAN: I think there's something bigger going on, and we talked about a lot of that today.

JOHN: Not a lot of people want to hear it, but it needs to be said.

Cultural Management *for* Organic Weed Control

KLAAS MARTENS

There's a real reductionist way of looking at nature that I think prevents us from seeing the whole truth sometimes. I had to step beyond that. In fact, one of the things that bothered me when we farmed conventionally was that if you took the model that I had learned—and the assumptions that I was farming under—the observations contradicted them, and it was enough to drive me crazy.

In the bibliography of an old ag text, I found a paper written by a German professor in the thirties, who was the pioneer in using chemicals to control wheat. He wrote that cultural practices form the basis of all weed control, while the various chemical means should be regarded as auxiliary only. It was an interesting comment coming from the pioneer of spraying weeds.

I started to study what these different weeds and pests do in the soil. That grew into a system of how to read what the soil is saying and how to understand the language that the fields are using to try to teach us.

I feel more comfortable if I don't put our rotation in stone. I let the weeds in the soil tell me which crop it wants to go to.

Modern agriculture has gone to being a series of materials—things that we can buy—that make us able to continue the bad behavior, the offending behavior. But we never deal with the actual underlying problem, which is why there's damage to the long-term productivity of the soil.

KLAAS MARTENS is a third-generation farmer who grows organic broadacre crops on over 1,600 acres in upstate New York. He owns and operates Lakeview Organic Grain in Penn Yan with his wife, Mary-Howell. Klaas has been very successful in developing organic farming systems to produce exceptional, high-quality yields of small grains.

Halfway through his transition to organic, Klaas was suddenly confronted with paralysis in his right arm—the result of exposure, he believes, to pesticides and herbicides. The arm took several months to heal, but it spawned an intense desire to learn about natural farming practices. This drove Klaas and Mary-Howell to pre-1945 texts—those written before the mass adoption of chemical agriculture.

Klaas is an expert in weed management—how to use cultural management practices, appropriate tillage, and crop rotations to manage perennial and annual weeds and to shift the weed populations of a farm over time. He has some very key insights into how to manage a number of different weed species.

JOHN: Klaas, can you describe your path from conventional farming to regenerative practices?

KLAAS: This has been quite a quite a journey for us in the last twenty-five years—to go from being fully conventional farmers who used all the chemicals to completely convinced organic farmers. We're not unique in that; I think you've had a similar experience, haven't you?

JOHN: Well, certainly the farm that I grew up on was very mainstream, using very intense pesticide applications. The farm is not organically certified, but there is no insecticide or fungicide or herbicide use at all—no synthetic pesticides of any kind.

KLAAS: It's a real journey in learning, and you wouldn't have had the handicap that I did—if we can call it that—of having gone through the land grant university system, learning how to farm "correctly," and then having to unlearn a lot before things started to work.

JOHN: Several of the mentors that I've worked with have told me that I would not have been able to learn what I've learned so far in the speed that I learned it with a university education. I don't have the context to be able to comment on that, but I found that to be an intriguing observation for sure.

KLAAS: I found it to be both a handicap and help. I found that there are people in the university who are good, honest scientists who have offered me a huge amount of help in figuring things out. The handicap is more in what we were being taught and how we were being taught to think and look at the farm.

There's a real reductionist way of looking at nature that I think prevents us from seeing the whole truth sometimes. I had to step beyond that. In fact, one of the things that bothered me when we farmed conventionally was that if you took the model that I had learned—and the assumptions that I was farming under—the observations contradicted them, and it was enough to drive me crazy. I either had to discount my observations or question my assumptions. For a lot of people, it's actually not very comfortable to question your assumptions.

JOHN: Dr. William Albrecht, at the University of Missouri back in the fifties, said, "Read books and study nature. And when the two don't agree, throw away the books."

KLAAS: Even there, I think there was some truth in the books, but not complete understanding. We have to figure out what part of it is useful and which part is not. This asks more of us than just reading it back.

JOHN: Klaas you've had a very interesting pathway and have done some fascinating work on your farm. What have been some of the memorable moments that led you to where you are today?

KLAAS: One happened at the Cornell University library. When we first converted to organic, like a lot of people, we were looking for what organically approved inputs we could substitute for our fertilizers and which organically approved techniques or machines or whatever we could substitute for our herbicides. In searching the library stacks, it became obvious that we were looking for weed control and fertility books from before 1945. It seemed like there weren't any until my wife suggested that maybe we should look in the archives—in the stacks, the older stuff that had been put away for storage. And there I ran into what had been my ag teacher's old text.

In the bibliography of that book, I found a paper written by a German professor in the thirties, who was the pioneer in using chemicals to control wheat. He wrote that cultural practices form the basis of all weed control, while the various chemical means should be regarded as auxiliary only. It was an interesting comment coming from the pioneer of spraying weeds.

JOHN: We've certainly shifted from that paradigm substantially.

KLAAS: Yes, we have. Here was the visionary who started a practice actually telling his students not to rely on his methods. That tells me he saw a lot of things that people since then haven't seen, and it turned my worldview upside down.

This is a practice that I learned is always reactive: you identify the weeds and then you figure out which poison will kill them—and hopefully not hurt the crop as much as the weeds. But he was telling us to look at the cultural practice.

Cultural practices are everything you do in the tending of the field—how you set up the situation that the crop is growing in, and the idea that what you did in the field has an impact on the field this year. This was radical and new to me. Rotation is the obvious idea. In this paper, though, he went through all kinds of examples of different practices that could impact what kind of species were there. The paper also inspired something that I had seen and observed—what happens when you abandon a field.

The first year after you abandon a field that's been in row crops—let's say it's been a corn field—think about what weeds will grow. Obviously, there'll be lamb's-quarter, pigweed, foxtail, velvetleaf, the whole range. They're mostly seeds of weeds that make huge numbers of seeds. You may end up with millions of seeds per square foot on the field. But what grows the second year? Now from our reductionist way of thinking, we would assume that because we just made that many seeds, we should have a lot bigger problem with those weeds. But the second year, none of those plants are growing. We have other weeds growing there. If you take that forward several more years, you start seeing goldenrod, woody plants, brambles, sumac—you know, all the thorny stuff and all the multiflora rose. And if you take it forward a few more years, you're going back to forest.

This is something Dr. Albrecht wrote about. If you let it go for five hundred years, at least where we are, it would be back to old-growth hardwood forest—mostly oak—which Dr. Albrecht called the climax crop. That's the kind of a steady state that nature would put the land in. This is how the land was made to work: it creates this succession where all of the species and communities—each one—changed the soil. The reason all of those weeds that set seed didn't grow the second year was that the plants that made them changed the soil, and that the right conditions weren't there for those plants to grow the next year. So other plants grew. And that group, again, changed the soil, so that another group grew.

Thinking back to what Dr. Albrecht wrote, he talked about these successions as actually being more productive, more diverse, and more vibrant than the climax crops. So that the tallgrass prairie and the oak would have been very stable, very resistant to invasions and diseases. Those plants didn't get sick—they could tolerate flood, storms, whatever, and remain in very healthy condition. Albrecht used to tell his students to see what they were looking at—to see how nature does its crop rotation.

JOHN: Why did you look for books that were written prior to 1945?

KLAAS: My father used to laugh at me when I said we needed all those chemicals. He said, "How do you think people fed themselves for most of history? These things are all recent." I just reasoned that if I looked back at those agriculture books from before 1945, I might be able to learn how people farmed back then, and how they actually made things work. That was my reductionist thinking to some extent, but I was reaching back to older knowledge. We've lost a couple generations of people who had passed all this knowledge down from one generation to the next—it all suddenly became irrelevant when the chemicals came along. I didn't know what else to do. I was in a situation.

Partway through our transition, I went to pull up the sprayer and found that my right arm was paralyzed—it wouldn't move. And I spent the summer paralyzed. I went to a naturopath and a chiropractor, who finally restored the use of my arm. We had a decision to make—at this point, hiring somebody to do a job that I figured would probably kill me or harm me in some way was not something that, as a Christian, I could do. There's something morally sick about hiring someone and telling them to do a job that you're convinced is going to be harmful.

That placed me in a spot: we wanted to do this for a living, but we had to figure out how to do it without using those chemicals. That puts a little more perspective on our visit to the library. We were desperately trying to figure out how to remain farmers but to do it in a different way.

JOHN: From the research that you've done and the description that you've just shared about plant succession in developing soils, how have you found that useful for managing crops and for managing weeds on your operation?

KLAAS: I mentioned that there was a conflict between my observations and our assumptions. All of these observations—when I think about them correctly—were pointing to a system that was very carefully engineered with an intelligent design behind it. It worked according to a logic and an order that was not random. I had to shift from thinking that these are all random occurrences that we're seeing—that these are the result of an order. Looking at the world that way, we suddenly realized that this is all cause and effect: what we do today, we will see the results of later. This really helped us in starting to understand what we had seen.

A mentor told me, "If you want to get the right answers, you have to figure out the right questions." So instead of seeing my field of velvetleaf and asking how to kill it—which is what I had learned when I saw these events as random—I needed to ask why it was there. What's more, what was it doing? What did the Creator put this here to do? Why did he put it here? And what was its function? I could keep on coming up with these questions that I had never thought about, but they are important.

I started looking at a pest that forced me to start asking why it was there and what exactly it was doing in the soil. Take this back to the succession that we observe. Obviously, these plants are changing the soil, and left to their own devices, they kind of work themselves out of a job and something else grows. I had to look at everything that I could observe. I had to try to see everything there was to see—look at it through new eyes—look at it through something that is working exactly as it was intended to.

The problem was me. If I didn't like something, I had to own it and say, "This is the result of what I've done up until now. Now, how do I change that?" More importantly, how do I learn from it? I started to study what these different weeds and pests do in the soil. That grew into a system of how to read what the soil is saying and how to understand the language that the fields are using to try to teach us.

There was a weed that at one point I thought was going to make it impossible for us to farm organically. I was really frustrated. It seemed like this velvetleaf grew taller than the corn, no matter how carefully I cultivated—a lot of it always survived. I had a one-acre spot in particular where I ended up mowing it. The corn wasn't going to be a crop; there was nothing developing. After about three or four years, it wasn't quite as bad, and the area where the crop didn't amount to much was smaller. Fast forward another three or four years, and lo and behold, my velvetleaf was getting into midsummer, and then it was starting to turn yellow. The lower leaves were turning brown, the lowest leaves had fallen off, and before the end of the summer it was dead. Not only that, but instead of being taller than the corn, it was only about three to four feet tall.

I called a friend at Cornell who is a lead ecologist. I was still thinking completely wrong. I told him I had found a disease that was going to make me a millionaire. My brilliant idea was that we could catch those spores and make a product out of them. My friend came out and looked the situation over.

He said, "I'm familiar with these leaves, and you can go ahead with your plan. But before it can be successful, you need to explain this to me: why is it that when that disease is in your neighbor's field, on his velvetleaf, it doesn't hurt his velvetleaf?"

And sure enough, this disease existed right across the road, and it wasn't hurting the velvetleaf.

I should have been able to figure this out quicker than I did. But I have to admit, I was quite dense, and I needed quite a few lessons and to notice quite a few things before I started putting two and two together. The next thing we noticed was a second disease in that velvetleaf that a student at Cornell identified as a virus. In the meantime, because I was paying so much attention to thinking that this was going to be my new product, I noticed that those leaves were covered with white flecks. The first time I saw it, I crawled on the ground and I said, "Look at all these white flecks—my leaves are just being eaten alive."

And the agronomist said, "You better watch out—you're not going to have a crop left with all these insects out here."

But then we looked at the corn and there aren't any bugs on the corn. The corn was perfectly healthy and growing well; it was only the weeds that had bugs on them.

The insects were actually carrying the virus, and the fungus was blowing on and killing them. But it wasn't this complex that was actually killing the plants—those were just opportunists. We had changed our system so that it had become a very unhealthy soil environment for the weeds. Because the weeds were unhealthy, all these pests were moving in and were attacking the weeds. It wasn't really the pests that killed the weeds; the pests were just there because the weeds were so sick that they weren't fit to live.

JOHN: We've actually observed a very similar situation on farms that we've worked on. A farm that we had worked with for a couple years had major challenges with Canadian thistles. After a few years of being in a small-grain rotation—a forward rotation planted back into a corn crop—we went back into an organic corn crop. No herbicides were applied, and it was cultivated. The cultivator got most of the Canadian thistle, but there were still Canadian thistles that were left right in the row. By the time the corn was tasseling, all the Canadian thistles were dead. They were brown. We sent those into a laboratory to try to identify if there was a pathogen or disease that had infected them. We weren't able to find anything on the dead plants, but it was quite an interesting experience.

KLAAS: How did you interpret what you were seeing at that point? Did you figure out right away what was going on?

JOHN: Our interpretation was that we had changed the biological profile of the soil biology to such a degree that the rhizomes were no longer able to survive, and that the plant ended up dying. The really interesting part was that it wasn't just the upper parts of the plant that were dead. The entire root system—and you know how ubiquitous a Canadian thistle root system can be—it was dead, completely decomposed.

KLAAS: I have a friend whose great-uncle researched Canadian thistle in Germany at the university, and he discovered an organism that lives on the deeper roots of the Canadian thistle, and the thistle needs this organism to survive. It's a symbiont, and this particular organism can only live in anaerobic soil. Oxygen is deadly to it, and that's probably what you saw. The subsoil was compacted, and Canadian thistle tends to start in the wet spots. The low spots would be anaerobic, and when it had started and if it had everything it needed, it would grow, and it would spread.

There's also a poem. Farmers used to say,

Plant them in May, they'll be back in a day.
Plant them in June, they'll be back soon.
Plant them in July, and tell them goodbye, Canada thistle.

That fits what an old farmer mentor of mine once told me. He said, "I'm going to do something here, and I want you to pay attention because you're going to learn something." Just about the time those thistles were in full bloom, he took my mower and headed out there and mowed them down. That was in July. They come in full bloom in July. And he said, "They're gone now, and they won't come back." And he was right.

That fits what we're talking about because those big roots in July, when the plant's blooming had shrunk—that's when the most energy is in the top—because they've shrunk, they created air passages that brought oxygen down. When we cut the tops, they shrunk even more to try to refill their tops, and that did away with the anaerobic conditions in the subsoil.

JOHN: Fascinating. What do you attribute that shift and that success to? In other words, if you were to give advice to other growers around the country, how can you replicate that success? What is the system or the process to follow to achieve those kinds of results? And perhaps not just with velvetleaf but possibly also with other weed species?

KLAAS: We're asking that question, and I told this story to a graduate class at Cornell, and the students in this class are hoping they can help answer this question exactly like you're asking it. They've come out to our farm, and they've collected soil that they've transplanted to their greenhouse now and planted velvetleaf in. They collected soil from a conventional farm that was suffering from really bad velvetleaf and they're putting it in the greenhouse under the same conditions. They're going to do microbiological assessments in every measurement that they can think of, hoping that they can reproduce these results in the greenhouse side by side—to start explaining to us on a very basic scientific level what's going on.

The bigger picture—what I'm attributing this to—is that I think the soil has a different biology in it. I know that the velvetleaf was really reduced at the same time as my pigweed and my lamb's-quarter. These are all nonmycorrhizal weeds. This is my guess talking. I'm hoping the university gives us a true answer and can prove it. But my guess is that the mycorrhizae on our crop roots are actually attacking the nonmycorrhizal weeds or somehow going after them to help the crop compete with them.

There's a researcher from the University of Wisconsin who had similar observations. In some of his plots, he had observed a 90-plus percent reduction in the growth of nonmycorrhizal weeds, and he was guessing this was due to the mycorrhizae on the crop plant. But he said he couldn't reproduce it because he was doing it in different soils, and it seemed that the mycorrhizae were only able to attack and throttle the growth of the nonmycorrhizal plants in certain soils. My guess is that it was the healthier soils, which were friendly with the mycorrhizae, that were giving him this extreme depression of weeds, and the unhealthier soils weren't.

JOHN: Working with many different farms, many different types of environments, and many ecological regions, it's been hard for us to identify consistent themes. The way that I've usually presented this discussion is to say that as you shift soil biology, weed pressure shifts and weed populations shift. I would say that 90-plus percent of the farms that we've worked on—within the first two to three years, their weed population has shifted. What was once a dominant—a major weed species—is now subordinate, and it has been displaced by another weed species entirely.

KLAAS: You're changing the microbiology of the soil. I'm trying to figure out exactly how we're changing it and what works in our favor. I'm sure you and many of your clients have figured out that you can benefit from it, whether you know how it works or not.

JOHN: I'd like to go back and ask you that question. You perhaps haven't figured out exactly the causal mechanisms of the effect that you had on the velvetleaf, but what were the practices that you were doing that you believe led to this result and to this outcome?

KLAAS: WE were going to a much more diverse rotation, and we had started putting cover crops on everything—we tried not to have the soil uncovered over winter. We were plowing a lot more green matter in. But as much as anything, this diverse rotation changed our soil.

One of my early epiphanies was that after we had started converting to organic, I was plowing a field, and I smelled something that I hadn't smelled since I was a little boy. I used to love to ride with my father when he plowed. The soil had this powerful smell that I just absolutely loved. I have since learned that scientists have discovered that smelling this material actually makes you feel happy. It has a psychological effect on people.

JOHN: Fascinating.

KLAAS: It's called geosmin. You know the smell; I don't have to describe it, do I?

JOHN: Yeah, I know the smell. I know exactly what you're talking about.

KLAAS: THAT smell came back. That was actually my one of my first big observations. I'm pretty sure that smell is a vital sign that can tell you that your soil is heading toward healthy—that it's healthier than it was. If you don't have that smell, it should be a warning sign that something is not working right.

JOHN: I completely agree with you. That's actually an indicator that we use when we're scouting in a farmer's field—we want to know what the soil smells like. It's incredible how fast it can change. I think that perhaps the fastest that I've personally experienced was working with a strawberry crop in California that was in relatively sandy soil, so the biology could change fairly quickly. That soil went from having no smell—like it was the equivalent of a sandbox—to having that rich, earthy smell in two weeks. It was almost unbelievable.

KLAAS: Biology is so powerful and can change so fast. I think one thing we're missing is that we're only looking above ground because that's what we see. But we should be looking below ground. We're also looking at the big things, but we should be looking at microscopic things. It's harder to see them, so out of sight, out of mind—I'm afraid this is something that hampers us.

I'm going to recommend an *Acres U.S.A.* book. Maybe you've read it. It's called *Healthy Soils, Sick Soils*. It's a translation from German—a work done by an Austrian researcher named Franz Sekera back in the forties and fifties. I think it was published in the sixties by his widow, Margareth Sekera. He took soil and put it on a dissecting microscope, trying very painstakingly to see what he could see. He used stains and what, at that time, were the most advanced techniques.

He found an entire ecosystem in the soil and in these pores that were in the healthy soil. He discovered that the organisms in it were all aquatic—they were all water-living things. There was this whole world of thousands—possibly millions—of different species. What's more, they were being fed by the exudates from the crop roots. That's one of the mechanisms that I think the crop uses: orchestrating biology by the type of food it's pumping out of its roots.

Think about how people say that tithing is difficult—when you use some of your bounty to give back. Think about plants who are giving half as opposed to 10 percent! And think about the number of species that these plants are cultivating around their roots and what they must be doing.

JOHN: Absolutely. Horst Marschner and Petra Marschner, in the third edition of *Soil Nutrition in Higher Plants*, refer to some plants that, in some conditions, can actually contribute as much as 70 percent of their total photosynthetic production back to the rhizosphere as root exudates. That's an incredible number when you stop to think about it.

KLAAS: Yes, and it's not foolishness on the part of the plants, this is giving back. These organisms are what builds the soil. Another thing that was in this book by Sekera was that they took soil that had good aggregate stability and put it in the lab at 70 degrees. Within two months, the aggregate stability had broken down.

JOHN: That's incredible. Breaking down at 70 degrees Fahrenheit is incredible when you consider that many, many soils during the growing season are exposed to the sunlight, and in many cases, the top couple inches substantially exceed 70 degrees—sometimes it's as much as 120 degrees or greater.

KLAAS: Exactly. What Sekera was saying was that these organisms were physically what was building the soil—the structure—but they needed the sustenance from the plants. At seventy degrees, starvation set in, and these microbes couldn't hold the soil together anymore. They did the same experiment with pesticides dumped into the soil, and the soil fell apart within days.

The book talks about tillage and about the building and maintenance of crumb structure. And it gives some rules for tillage. It says you should never disturb a soil deeper than what you've already developed a healthy crumb structure to because, below that layer, it can't rebuild.

JOHN: Going back to our conversation about weeds and cultural management of an ecosystem and how that might lead to shifting weed populations, there is also an interaction, if you will, between weeds and the crop that we are trying to produce. I know that in a previous conversation, you mentioned an experience that you had producing dry beans, and mustard was present in the fields. Can you tell us a little bit about that?

KLAAS: When we first converted to organic, one of the things we noticed was that we suddenly had customers wanting to buy almost anything we could grow, which was quite a switch from taking corn to the elevator and seeing how bad the price was going to be and if we could still make any money at it. When we went to organic, it was more or less, "What would you charge me to grow this?" or "What would I have to pay you if I could buy some beans?" Growing dry beans was almost like printing money at first—they were so profitable, and the yields were terrific. I was very happy with them.

But after about the second time that I grew them—even after five or six years in between when I planted them again—I decided I must have gotten bad seed or that I had gotten a new variety of seed that wasn't as good. They weren't growing as well. By the third time that I grew these beans, the roots rotted off, and the yield had collapsed, and the crop was really awful.

I went to Cornell again, and my old friend there, George Abawi, was near retirement. He had studied nematodes and roots and fungi. He said, "You know, I've been in this business all my life, and when these farmers came to me with a problem, I always wanted to explain to them what was causing the problem. But they always said 'No, no, no—don't bother with that. Just tell me what to put on it to kill these things.' And finally—now that I'm ready to retire—someone's actually asking me what's going on." He was very happy to help us.

He explained to us that there were root fungi pathogens that were carried by this crop. But it might have been five or six years since we had grown dry beans. He asked what else we had planted in between. I told him that we were growing red clover, and that we would seed our small grains down to red clover, and we were growing soybeans, and we had this really nice, healthy, diverse rotation.

He said, "Virtually every crop you're growing either leaves the problem the same or makes it worse, so that by the time you're back into dry beans, you've multiplied the amount of disease pressure that's in there because not one of your crops that you're growing in that rotation actually reduces these pathogens in the soil. They all just promote them."

He developed an interesting experiment. He asked the question, "How will the previous crop affect the root rot complex in dry beans?" He picked a very susceptible variety of dry beans, one that had no root rot resistance. And he took a field that was fouled with the disease, and he divided it into squares. He planted about twenty different crops in it randomly. The next year, he planted these susceptible dry beans in the whole field, and he noticed that in some of the plots, the dry beans looked great. They were growing and thriving. But there were other plots where the dry beans actually died because the roots had totally rotted off.

Then he started asking which crops were the best previous crops for the dry beans. He noticed that some crops made the problem a lot worse, some made it a little worse, some were neutral, some helped it a little, some helped it a lot, and a couple of them devastated the disease complex and produced a very healthy harvest of dry beans. The best one was yellow mustard.

That's a talk in itself—all the mechanisms of the yellow mustard that kill this dry bean disease. But we also realized later that it wasn't just a disease that was in the ground; there were nematodes. The nematodes were breaking the roots and chewing them up, which allowed openings for the disease to move in, which is why the dry beans had such terrible roots.

As a side note, when I was studying this problem, I found some pictures of dry beans from back when the Native Americans grew them in their system, and some of those roots were four feet deep. I couldn't imagine dry beans with roots four feet deep; the ones in my field had just rotted off and were about two inches deep. This has to do with the health of the soil.

So we added yellow mustard to the rotation. We discovered it'll grow at low temperatures, so we threw it in our corn—where we had grown corn the year before—and let it grow until early June, and then we turned it under ahead of the dry beans. It worked, and it actually worked sufficiently to make quite a difference.

But George also said not to do this everywhere. He said to leave a few spots in the field and mark them where we didn't have mustard. He came back that summer and evaluated the roots, and he could tell me with a shovel where I'd had mustard and where I hadn't—just by how the roots looked on the dry beans. There was a dramatic difference. By adding one new species—the right one—we had eliminated this whole problem. Our whole system hadn't had any brassicas in it.

JOHN: How did your weed profile change once you incorporated brassicas into your cover crop rotation?

KLAAS: We had been seeing an increase in mustard growing in our fields. I'd never had mustard problems. At the same time that these dry beans were failing on me, we were having mustard pop up in the corn. Mustard is relatively easy to kill with a weeder, and it's relatively easy to cultivate out, except that you've got to be on the ball. It's very hard to get rid of if you're not there in the right window.

I should have noticed that—this all goes back to seeing what I'm looking at. I was seeing this problem in the dry beans, and I was seeing this annoying increase in mustard. If I had been paying attention to my soil, I would have said, "Hey, that soil is trying to tell me something. It's trying to bring forth mustard." Actually, looking back, the soil was trying to heal itself—from the sickness that I was putting into it with an unbalanced rotation—by bringing up this weed.

I told this story some years ago in South Dakota, and one of the farmers popped up and said, "You just explained one of my biggest mysteries. I had this field that was coming up with mustard, and it kept getting worse. The last time I planted oats, there was such bad mustard that I just let it go to bloom and plowed it under because there wasn't going to be an oat crop anyway." He said, "You know, I haven't seen mustard in that field since."

JOHN: It corrected the problem that it was designed to correct.

KLAAS: That's right.

JOHN: Do you still have mustard in your field since you've incorporated it into your rotation?

KLAAS: No, and I've been throwing the mustard on, and it doesn't even grow real well. I haven't seen it come back as a weed since.

JOHN: Klaas, you've mentioned that you have a very diverse crop rotation. What does your crop rotation look like?

KLAAS: At this point, we have twenty to thirty different crops, and it's very flexible. I can drive my son and some other people I work with crazy because of the changes that I make, but I like to be able to respond to what the soil is showing me. There's a basic rotation that we had started with that was very profitable, but we kind of wore it out. That was corn, followed by soybeans, followed by winter spelt, underseeded to clover, and then the clover turned under the next year for corn. That rotation was used by the university here and by most of our neighbors.

The problem is that after about ten years of this rotation, that everyone started having the same problems. There was an increase in Canada thistle on some farms; there was this increase in disease. Soybeans are much more resistant to the root rot in the nematodes than dry beans. But if somebody were to substitute dry beans for some of the soybeans in that cycle—and that's what we were doing—they would have problems.

The first thing we added to this was a second cover crop: the yellow mustard in the corn. But then we were starting to have more trouble, and the clover wasn't growing as well. We started seeing Canada thistle here and there. In the worst fields, we went from the winter grain spelt to a second winter grain—a winter barley. We double-cropped that winter barley with buckwheat the first couple of years. By double-cropping we were breaking the ground up in midsummer, in July, when you can hurt the Canada thistle and the other perennials. We weren't doing a lot of tillage. We were doing a deep tillage—a deep ripping—trying to leave the mulch on top. But then we discovered that buckwheat is not the most profitable crop in the world, and changing this rotation helped our soil quite a lot.

We discovered that we could plant dry beans—pinto beans—after the barley and then go back to a winter grain and then start over. That stretched it out to a five-year rotation; there must be eight different species grown in that rotation. From there, we've diversified further. We brought in vegetables after the clover so that we're not growing corn quite as often. I would plant cabbage, and we've thrown in squash after the small grains or even after the corn sometimes. From that, we've added flax, we've added oats, we've added barley. Instead of just spelt, we've got rye, winter barley, spring barley, and oats. We planted kale. It seems like the fine-tuning of this basic rotation—I feel more comfortable if I don't put it in stone. I let the weeds in the soil tell me which crop it wants to go to.

It's still this old, basic framework that can branch out in a hundred different directions, and I'm letting the weeds and the soil direct me as to what I do with it. Sorry, that's not a very satisfying answer for people who want a set program or a formula, but I think the bigger answer is that we need to be open to what the soil is showing us and be more flexible in our rotations. Think first of what will make the soil healthy and secondly of what's bringing the highest price.

JOHN: Don Huber once told me that 85 percent of the benefit from crop rotation or from cover crops occurs from the immediately prior crop—the crop that was planted just before the crop that you are growing. Your rotation depends on what the soil and the weeds are telling you. What are the variables that you're monitoring? And how do you choose and select which crop you're putting based on what you're observing in the field? What is the soil telling you based on some things that you look at?

KLAAS: I look for a field that is dominated by a certain color. This is something that biodynamic farmers taught us years ago. Yellow blossoms point to sulfur. The compounds in the yellow mustard that have the big activity are sulfur compounds. Also, when we looked at our soil tests, where we had mustard growing, our sulfur was tied up—we were deficient in sulfur. Even though we thought we were putting plenty of sulfur on, those weeds were telling us, "No, you're not—you're not putting enough on. We're still short."

The blue flowers are quite often a phosphorus indicator. White flowers indicate potassium, but they can point to calcium, depending on the situation. If I see corn chamomile overrunning a field, I can be quite sure that there's a calcium issue, and if I see milkweed, I'm 90 percent sure we're short of zinc. There might be zinc on the soil test, but the crops aren't getting it—it could be complexed. Hemp dogbane is the same.

I don't want people to do this instead of soil testing. These are little data points and observations that I've made that are parts of the conversation—when the soil is showing me things.

I have to credit Dave Maddox on the zinc. He drove by a couple of my fields, and he said, "I see you have a zinc deficiency."

I said, "What are you talking about, Dave? You haven't done a soil test here yet."

He said, "I don't need a soil test. I see your milkweed."

I thought that was an interesting little anecdote. But if I see foxtail, I can guarantee you that the soil is suffocating, that it's not getting enough oxygen, and that it's not able to get rid of the carbon dioxide and the ethylene and some of the products from respiration that build up. I used to tell people it's tight soil, but it's actually more than tight. You could have pure sand, and if you load it with foxtail and have enough rain, that soil will not breathe.

If I see foxtail growing, and if I've got an opportunity in the rotation, I would love to grow millet because millet loves the same conditions that foxtail does. Genetically, it's very similar to foxtail and will repair the same maladies in the soil. If I see the soil trying to grow foxtail, I'm going to have some millet there the first chance I get because I know the millet will love that soil condition, and the soil will be corrected by the millet.

If we see certain weeds that either have a given function or that are closely related to a crop, that's the soil telling me that this crop is going to do great here. Don't try to plant what you were planning on—the soil is showing you with its weeds that it's not going to be happy. I could go on and on with examples, but that's where I'm coming from. That's how I figure these things out.

JOHN: That's a that's a great answer, Klaas. Thank you. Essentially, what you're describing is managing an ecosystem: looking at the entire soil-plant system and managing how it's developing and evolving with crop rotations and with tillage and with the various tools that we have. How have you seen the ecosystem evolve in regard to disease and insect resistance as you looked at managing the entire ecosystem? I know that in earlier discussions, you've referenced a couple of examples of managing aphids on lettuce by managing the ecosystem, and I wondered if you would share that.

KLAAS: The aphids on lettuce are one of those absolutely amazing examples. This was university research in California.[5] There was an aphid that was getting into the heart of the romaine lettuce and making it totally unmarketable. One of my friends there was doing research on how to control this in organic systems. Anyone who has been to the Salinas Valley and some of those parts of California knows that these are vast wastelands of only one species—only lettuce. They found out that the sweet alyssum flower promotes an insect that controls the aphid. It actually eats the aphid. They found out that if they had one sweet alyssum planted in every thousand square feet throughout the lettuce field, that this did not reduce the yield of lettuce at all. The yield was identical, but that one plant every thousand square feet completely

5 Eric B. Brennan, "Agronomic Aspects of Strip Intercropping Lettuce with Alyssum for Biological Control of Aphids," *Biological Control* 65, no. 3 (June 2013): 302–11, https://doi.org/10.1016/j.biocontrol.2013.03.017.

eliminated the aphids from the field. That's an example of how introducing the right plant at the right time and in the right place can completely remove one group of insects.

JOHN: That's almost nothing. That's forty-four plants per acre.

KLAAS: That's right. To have that dramatic an impact—I'll bet the cost of the spray they would have been putting on every week or every ten days would have far exceeded the cost of those flowers.

JOHN: What have you observed on your own farm in regard to disease and insect resistance over the years?

KLAAS: We're paying attention to Brix. When we have healthy plants, the bugs just don't seem to hit them. Although I have seen a cabbage aphid in our cabbage. Canadian aphids must be a special case. They drove me crazy at first. Then I noticed that where I had corn planted next to cabbage—as long as the strips are not more than 150 feet wide—I had no problems with cabbage aphids. We learned afterward that the lady beetles were feeding on insects in the corn. About the time the corn matured to the point where there was no food for the lady beetles, they just hopped down into the cabbage and started cleaning up the aphids in the cabbage field.

That was very similar to the mechanism or mode of action that we saw in the lettuce. I'm wondering now, after the lettuce experiment, if somebody could just drop some sweet corn plants out in their cabbage field. That might be insurance against the aphid, but having the narrow strips certainly works well too, and that saves you having to cut roadways open. You've got a different crop right next to it.

Another thing we've seen in our fields is that if we start seeing leafhoppers moving into either dry beans or alfalfa—or potatoes or any of these crops that have leafhopper problems—I'm almost assured that these plants are low in magnesium. I found that I can spray Epsom salts on the field instead of an insecticide, and the crop will start greening up and perking up and looking better. In some cases, the insects totally disappear. They just avoid those plants.

JOHN: I have a thought for you on the both aphids and the leafhoppers. It's been our experience working with a number of different crops that there's this triad, if you will, these three different nutrients: magnesium, sulfur, and molybdenum. Each of those three needs to be in adequate supply. If you have adequate levels of two, but the third one is low, you have challenges with aphids, for example, and leafhoppers. But if all three are adequate, those insect pests just disappear, and it's pretty profound. Actually, I would say we have a 90-plus percent success rate. I'm saying that just to be conservative. In reality, it's probably close to a 100 percent success rate dealing with this on a broad range of different crops.

KLAAS: In that case, our Epsom salts are covering two of those basics.

We may or may not know what we're doing, but we get the right results—at least on our farm. That's really an interesting point for you to bring up because magnesium and sulfur are quite often linked when we overfertilize or when we don't have the right fertilizer program. Being too high in magnesium will tie up magnesium; you can be either too high or too low in magnesium in the soil and end up with a deficiency in the plant. Or if you're short on sulfur in the soil, you could also end up being deficient in magnesium and calcium because they would become insoluble because of the situation.

JOHN: Exactly.

KLAAS: The big question isn't what the missing element is; it's really figuring out why it's missing. We're having to kill the bugs so that we have a harvestable crop, but we really should be addressing the reason that we had to go in and do something.

JOHN: The reality—and I'm sure you've experienced this as well—is that these full-year nutrient applications can be incredibly effective in pest removal. Not as a pesticide—nutrition doesn't contain any toxins of any kind—but we've had a number of experiences in which we put on a nutritional application, and spider mites were dead in forty-eight hours. Aphids and leafhoppers disappeared in twenty-four hours. It's pretty profound that it happens really fast.

KLAAS: And if the plant were able to launch a defense, it had been crippled by the shortage of that nutrient.

JOHN: Exactly.

KLAAS: These plants were created to be able to handle the pest. The pest and the plant were both created by the same Creator, and one can't live without the other; there would be an imbalance.

We also really need to pay attention to why the deficiency was there. We tend to see the soybean aphid in spots that were very high in potassium. High potassium tends to lock out magnesium. We can sometimes have deficiencies that are induced by excess. I've seen more farmers buy themselves expensive problems and expensive solutions. When somebody comes to sell you a huge amount of something—especially if it's out of the ordinary—be careful. You may be buying yourself a different problem.

JOHN: Klaas, you just mentioned a very key phrase—that excesses may actually be more of a problem than deficiencies. We first began experimenting with the use of sap analysis in 2011. We started using it on a very intense and broad scale in 2013. Now, after having several years of experience with sap analysis, I can say that most of the nutritional imbalances that crops experience are not a result of deficiencies. They are a result of the excess products that growers apply, creating deficiencies of something else. Ninety percent of nutrient imbalances are created by the grower. It's absolutely incredible. Many of the recommendations that we make consist of discontinuing product applications or positioning them at a different time. It's really profound.

KLAAS: You've put your finger on the biggest and most common problem that I've run into: overapplication of something that creates an imbalance. Von Liebig described the law of the minimum, and there's this illustration of the barrel—with which stave is lowest. But then André Voisin wrote a pamphlet about the law of the maximum, in which he said that when any one mineral is in excess, it can cause a shortage of one or more others.

JOHN: Exactly. Klaas, you think a lot about what is happening in agricultural ecosystems from a macro perspective. What is something that you believe to be true about modern agriculture that many other people do not believe to be true?

KLAAS: To me, modern agriculture—and I'm going to include genetic engineering—is a series of enabling tools that allows farmers to use bad management. In a natural system or a balanced system, where farmers are in harmony with the creation—with the soil or their farm—when they're doing something they shouldn't, the soil is basically screaming. If we persist in doing it, the soil starts screaming very loudly, and we see these problems pop up—be it a weed, be it an insect, be it a disease, whatever.

These are symptoms. Modern agriculture has gone to being a series of materials—things that we can buy—that make us able to continue the bad behavior, the offending behavior. But we never deal with the actual underlying problem, which is why there's damage to the long-term productivity of the soil.

It's a little bit like the idiot light coming on in your car, telling you there's low oil pressure: putting tape over the light or clipping the wire takes care of the immediate symptom. The light goes out. But that doesn't deal with the reason that light was on. Or let's take it on a spiritual level. Let's say you're in error or in sin, and the minister or the bishops are pointing out that you really need to stop doing that. Well, we can ignore them, we cannot listen, we can go find another bishop who will say something different. But it doesn't take away the problem—it won't take away the long-term results of the problem.

JOHN: We become addicted to point-and-shoot solutions rather than identifying the underlying causes and really thinking about the underlying causes of various challenges.

KLAAS: Right. And those challenges, which are seen as our enemy instead of as nature, essentially are pointing out the error to us.

JOHN: Klaas, you do a lot of reading and a lot of thinking about what is happening on your operation, things that you see happening around the country. What are some books or resources that you would recommend to growers?

KLAAS: I love the Albrecht papers, but they can be dense. It can take a lot of discipline to learn from them. There was an old Cornell text on weeds that was used for about forty years. Conrad Muenscher was the author. It was simply titled *Weeds*. What I loved about it was the line drawings of all the weeds. And it gave control methods. For each one, it gave them in a community; for a family of plants, it would give all the members of the family. It would describe the differences between them and would give descriptions to help you identify them, but it would also state where they grew, when they grew, and what conditions favored them. Then it gave a list of cultural practices that could be used to eliminate or reduce them. Especially in my early years, I found that book to be really helpful because instead of just identifying the weed and finding a chemical or a tool to kill it, this was identifying the weed and teaching me about it, giving me information that helped me understand what the weed was telling me.

Another one that I would highly recommend everybody should have is André Voisin's *Soil, Grass and Cancer*. It's a really brilliant piece of observation that connected the minerals in the soil to some really bad diseases—both shortages and surpluses—diseases that modern medicine still can't get its finger around.

I caution against looking at this as written in stone. I think we all know a lot more than we give ourselves credit for. I'm hoping that the big thing I could teach people is to use what they already know—to use what they've already seen—and then to look forward and train ourselves to see more, to see the right things so that we can learn—so that we can draw our own conclusions about our own farm. Because nobody knows more about your farm than you do. That's the one that you live on. That's the one that you see every day, and you don't need an intercessor to talk to your soil. You can talk to your soil yourself and learn from it and pray for the Spirit to guide you to see the right things. Also pray for the discernment to figure out the right message from all we're seeing.

JOHN: I love that. We are the experts on our own operation. Nobody knows it as well as we do. I think that's a very, very key thought that is often lost.

KLAAS: I think it's very important for us as advisers to remind ourselves that we are only a tool. I've been given gifts of observation and gifts of knowledge, but I'm not the source of that knowledge. I have to give credit to the source of the knowledge and be thankful that I can be the conductor of it—that I can share some of it with other people.

JOHN: What is the question you wish I would have asked?

KLAAS: The question would be, "Where do we find help and guidance and discernment in managing our farms?" Because I think that's really the most important question. I don't want anyone to come away thinking that any one person is the source of what's good.

JOHN: What sources have you found that are useful and helpful for guidance and input that are important for the growers that you are working with?

KLAAS: The most important source I have found has actually been having time—quiet time and prayer time in the field—asking for discernment and asking for help. But along with that, what was really important—and I'm sure you play this role often in your work with farmers—is to have farmers or people that can give guidance and discernment to help me.

For that I find groups of farmers. We have our group that meets every winter here in Geneva, New York. We can share observations, ideas, what worked, what didn't work. We need to keep each other grounded in doing this so that this becomes a community thing. But we always have to include in that community the source of all that knowledge and all that wisdom, which is not any one of us. I've been entrusted with a lot of insight and some really great gifts. But none of them came from me. All I can do is hope that I can steward them correctly.

CONCEPTS *in* BIOLOGICAL FARMING

GARY ZIMMER

I didn't create or develop too much. I just paid a lot of attention.

Soil health is about having a biological balance, a mineral balance, and a beautiful structure—loose, crumbly, chocolate cake structure. How do you achieve that? You have to have a variety of crops. You can't have a crust. You can't have packed ground. You can't overtill it.

I would probably say that the thing that's the hardest is getting biology to work. If biology is working, it provides structure.

I find it offensive when people say, "How are we going to feed all these people?" I say, "Forty percent of our corn goes into making ethanol, and we have the capability to double our yields. And we can do it with a cleaner method of farming."

If you truly want to sequester carbon, you have to back off of nitrogen because you're burning carbon up with nitrogen. If you don't back down on nitrogen, you're never going to build carbon.

THE BIOLOGY in the soil is undeniably the engine that drives healthy plants. The multitude of bacteria, fungi, nematodes, and other creatures that surround healthy plant roots provide nutrients and water to plants in the correct form, at the correct time, and in the correct amounts.

But how do we as growers get that biology to start working? This is the question that Gary Zimmer seeks to answer.

Gary is a dairy nutritionist and one of the original cofounders of Midwestern BioAg. He has been a leader in the biological farming field for several decades, through his consulting and speaking and as the author of *The Biological Farmer* and *Advancing Biological Farming.* He operates Otter Creek Organic Farm near Lone Rock, Wisconsin.

JOHN: Gary, you've done a lot of work over the years in many different areas, with your background in dairy nutrition and the work that you've done with Midwestern BioAg. What are some of the memorable moments that led you to where you are today?

GARY: I think there was a lot of questioning going on back in the seventies. I was teaching at a technical college, and I got introduced to the fact that what we put on the fields affected what we fed the cows. That was like, "Oh my gosh!" That meant that we could change feeds. Prior to that time, most dairy people—whatever they grew—they would just take a sample and add what was short or dilute it down. We realized that we could change that feed. We ought to be able to change food and a lot of things we consume through healthier soils and better mineral balance. That was one of the moments that really changed my career.

Then—teaching at that technical college—I was a dairy nutrition guy, but I had to teach soils and agronomy and farm finances, so I brought in outside speakers. I was introduced to Brookside Laboratory Farms out of New Knoxville, Ohio. Back in those days, they were a pretty big force in southeastern Minnesota, where I was. I learned an awful lot from some really good mentors. I always tell people, "I didn't create or develop too much. I just paid a lot of attention." As time goes on, we all got better at understanding soils and soil health. I had some really good teachers in the beginning through that Brookside experience.

JOHN: Gary, there's a lot of conversation around how we need healthy soils to produce healthy plants and to produce healthy food for people. I sometimes hear people saying that this link is weak. As a dairy nutritionist, how much can we impact food quality by impacting soil health? Does it really make a difference for animals? And by extension, do you believe it really makes a difference for people?

GARY: I think it starts with finding limiting mineral factors. Back in the old days, we were under the assumption that N was N, P was P, K was K, and pHs were fine. Then we began to realize that the sources of those minerals were important for soil health and soil fertility. It's the exchange of those nutrients. Calcium is really king. It's not having the right pH, but rather having the right available levels of calcium.

Soil health is about having a biological balance, a mineral balance, and a beautiful structure—loose, crumbly, chocolate cake structure. How do you achieve that? There are some guideline principles. Way back when I wrote my first book, we came up with the rules of biological farming, and now they have the rules of soil health. I think there's a lot of common sense—common ground. You have to have a variety of crops. You can't have a crust. You can't have packed ground. You can't overtill it. There are all kinds of factors that affect what makes that work.

The thing that's hardest for people to understand is this mineral thing. Just because you have minerals dumped on the soil does not mean that the plant gets them in the right proportion. Sometimes one interferes with another.

Having said all that, soil does really impact our cattle feed. The minerals coming through the feed we grew would produce 50 percent higher mineral uptake into our forages. Those minerals are much more biologically available to the animal than simply adding a stone. As a human, if I said, "I'm not getting enough nutrition and minerals, and I'm going to take all these supplements," well, are we designed to utilize those supplements like we are designed to utilize the stuff that should be in our food?

JOHN: Exactly. One of the things that we've observed over the last couple of years as we've started using a lot of sap analysis—speaking to your comments on mineral balance—is that in many cases, the nutrient imbalances that crops are experiencing, particularly nutrient deficiencies, are a direct result of the excesses of the products that growers apply.

GARY: Absolutely. There's no question about that. These guys who dump on lots of nitrogen or potassium—with that potassium comes a lot of chloride in their salt-form fertilizer. They are definitely interfering with calcium. It interferes with magnesium uptake.

You brought up sap analysis. Just because you got a soil test and some lab says everything is ideal—you and I both know that doesn't necessarily translate into the plant having it ideal. These guys who are driving these healthy, high-yielding crops, how are they paying attention to minerals? The levels that are determined by some lab were based on an average crop, and they're using a lot of plant-protective compounds. What would those numbers and ratios look like? I'm sure you've seen a lot of things in your testing and career—that it isn't quite as simple as just grabbing a soil test and putting something on.

JOHN: Our goal as farmers should not be to create a pretty-looking soil test. It needs to be to grow a really great crop. I think that's been the fallacy sometimes—that soil analyses are used as the goal rather than as a measuring stick for where we want to go.

GARY: That's why, in my career and in my books, I've tried to simplify things so people can understand. And that's why separating out soil correction from crop fertilizer, I think, is a really good thing for farmers to understand.

I had a call this morning from a producer, and he said, "Well, our soil test looks pretty good. Does that mean we don't do anything?"

I said, "Well, if that soil test was correct, that means you don't have anything major to correct. But there are things like boron, sulfur, and soluble calcium we still need to feed the crop."

Soil has a certain ability to dish out minerals. How do we provide all those extra things and with the excesses that we've put on?

JOHN: Gary, you mentioned the three foundational pieces: biology, minerals, and structure—managing soil biology, soil mineral balance, and soil structure. Which of those three have you found the most challenging to address and to resolve?

GARY: I think we do understand the biology today. We called my company Midwestern BioAg because I wanted to focus on biology. But what do we do about it?

I think the minerals are probably the easiest to manage. But we still have to decide how much to put on and from what source. Biology and tillage—the physical aspects—all kind of work together. If you feed the soil a decent amount of different kinds of carbons, then you have to determine, "Which kind of carbons? How do I manage them? Do I just leave them laying on top? Do I shallow-till them?" You get all these questions.

Progress sometimes is slow. I think there's a real struggle in farmers to find what their limiting factor is. Of course they're limited by budgets and time. I would probably say that the thing that's the hardest is getting biology to work. If biology is working, it provides structure. That's probably the hard part to understand.

JOHN: That's been our experience as well—that the biology has been the most difficult to get going. In my observation, it's actually become more difficult over the years. It's harder now to regenerate biology than it was ten years ago when I started. I wonder what the implications are of continuing herbicide and pesticide applications and whether they might be contributing to that challenge.

GARY: My family farms organically, and when we took on this land, there wasn't a weed growing on it. It had been heavily sprayed with chemicals for lots and lots of years. In the first year, there wasn't a weed out there. Then we grew some crops, and we put some fertility on. And of course, it seemed to be struggling a little bit. Then we took it through a transition to organic and grew some cover crops. But no weeds went to seed.

This fall, we grew a small grain crop, and then we put in a nice cover crop. Following it, the weeds were everywhere. It looks to me like the soil came alive, and those little seeds that had been laying out there, suppressed by chemistry, are now back trying to work. It was really an interesting thing. Those weed seeds didn't get dropped on there. I think that soil was so suppressed with chemistry that the biology wasn't functioning.

Whether weeds or plants, it came back alive. As a dairy nutritionist, when I started this business, we worked with a lot of small family dairy farmers. Many of them are gone now, which is sad to see. But a dairy farm—because of manure use and because of rotation—because of having forages—and because they weren't as heavy chemical users—they were already half biological when they started.

JOHN: It is amazing what forages and what having a bit more crop diversity can do in the system. Did you know that in the state of Iowa, anything that is not corn or soybeans is considered a specialty crop? Alfalfa is considered a specialty crop.

GARY: Wow. I just drove all the way out to Nebraska last week. There's nothing but corn and beans there, so they're probably all specialty crops.

JOHN: In the wealth and the depth of experience that you've had, what has been something that really surprised you?

GARY: It seems like I've been at this for thirty-five years, struggling away. Everyone and his brother wants to get into organic now because they want the money. But they don't understand the system.

I think that seeing NRCS get involved and really push hard on soil health—I wouldn't necessarily agree 100 percent with them, but at least they're bringing up the discussion of the value of soil health and water infiltration and all these things. That was a real surprise to me—to see that this would happen among conventional farmers—that they would be talking about soil health and would be asking some of the questions that we've been asking for many, many years.

Now I see it in my world. I see so many conventional farms. Maybe they didn't laugh at us, but they did think, "Well, where's the research behind it?" Now they're asking all kinds of really good questions and paying attention. That's been a real big surprise to me. I didn't expect to see that in my lifetime.

JOHN: We're seeing a tremendous shift as well and a tremendous desire for information.

Continuing with that thought, that you're surprised that people are inquiring—because of the different perspective that you've had in the past—could you elaborate a little bit on that? What is something that you believe to be true about modern agriculture that is very different from the mainstream view?

GARY: Way back when I was at Brookside, in 1976, I visited a farm in northern Illinois that was growing 375-bushel corn, with a public variety of corn from the University of Illinois, on forty-inch rows. You could stick your hand in the ground clear to your elbow. It was this researcher's pet project. Of course it got my attention.

In the nineties, in Iowa, there was somebody who grew 400- to 500-bushel corn. That really opened up the doors, I think, to say that there's much more potential. I started my business back in the early eighties by leasing a fertilizer facility in northern Illinois. But we did very little business in Illinois. The soils are so good—all you have to do is knife-in anhydrous ammonia. If you have seeds in the hopper, you're going to get 150-bushel corn.

Now they're paying $12,000 an acre for places like that and are chasing 300-bushel corn. So we actually built a new fertilizer facility in Illinois and hadn't done much business here prior to that. I like my strip-till farms for conventional use. A lot of that land is rented.

This is why I find it offensive when people say, "How are we going to feed all these people?" I say, "Forty percent of our corn goes into making ethanol, and we have the capability to double our yields. And we can do it with a cleaner method of farming." It's not that we can't produce more food. Most people don't even eat the corn and beans. We can do a lot of things if we want to feed these people. But I think that conventional agriculture, or whatever you want to call it—modern agriculture—is shifting to a biological farming system.

JOHN: And that is exciting.

GARY: It certainly is. There are a lot of questions out here. I was just out in Nebraska, and it was a little shocking to me. They did an organic field day, but those guys only wanted to get into organic for the money. The whole group was just thinking about shuffling resources. "Instead of buying an insecticide, let's buy an organic insecticide. Let's buy one that's certified."

They miss the whole point of soil health. But they'll get there eventually if they're going to make this work. If you're going to buy organic nitrogen, you're going to find out that whatever you get paid for the product you grow isn't going to be enough. You better figure out how to make the system work.

JOHN: Gary, there have been other stories about extraordinary historical yields. Don Huber describes how a team he was working with was producing 450-plus-bushel corn back in the late seventies. And you've just described 375-bushel corn forty years ago. What happened with that technology and with those ideas? Why did we lose all that knowledge and information over the last forty years?

GARY: I think what happened in that case was that articles in the paper came out that said that the guy put 400 pounds of nitrogen and 300 pounds of phosphorus on. They didn't really understand soil health back at the time. Like I said, you could stick your arm in the ground and clear your elbow.

The other thing was that they didn't understand that a lot of those inputs came from manure. It wasn't from commercial fertilizer. The guys that went at it and dumped on enough commercial fertilizer to match his numbers didn't get the yields because they missed the other factor—that the livestock manure contained a biological link, and there were also other trace minerals and things. It wasn't just N, P, and K.

I think that at that time, there was a lack of understanding as to what it takes to make this system work. That's why it took a lot of years. Can we ever achieve the results that these guys talk about without all the intervention of chemistry, and without a rotation and different varieties of plants to put out there? I don't really think so. I think we need cover crops in between. Or we need another crop.

I think that to change agriculture is really simple. We need to subsidize something besides corn and beans. If farmers could just grow a small grain and then put a cover crop in afterward, they'd be putting nutrients back into the carbon biological cycle. It's not just chemistry—it's how nutrients are delivered to a plant and in the correct ratios.

JOHN: Gary, if you were president for a day, and you had the capacity to influence change in agriculture, what are the things that you would change?

GARY: Well, the first thing I would do is I would tax nitrogen so high that they couldn't afford to use very much of it. That's a step in the right direction. The second thing, like I said, I would subsidize a small grain.

The people that are starving in the world eat wheat better than they eat corn. There are issues with wheat—the gluten issues—but I think we've done things to wheat that probably caused that. I'd subsidize some small grains to give the farmer that rotation and then tax nitrogen really high.

I wouldn't subsidize cover crops, but I would subsidize food that grows. If you give farmers money to grow a cover crop, they never really figure out the system. They'd have to figure out how to grow nitrogen; they'd have to figure out how to get the plants healthier. If I were president for a day, I'd keep it simple. I don't think it's that hard to change agriculture. It's just what we put our focus on.

JOHN: Exactly. Gary, you have a lot of experience with boots on the ground, working with many different farmers from all over the world. Often, I find that there are a few key success factors on a farm that growers often overlook or may not be paying attention to. Do you have any insights on this, having observed so many different operations? What would you say are some of the often-overlooked factors to success?

GARY: I just started on a big project in Eastern Europe. The fields over there that have fairly high levels of calcium to begin with have a real benefit compared to those that start with a high-magnesium/low-calcium soil, or a high-potassium or high-sodium soil.

Also, in Eastern Europe, farmers probably grow six or seven different crops. They have a variety of different crops. Some of them are taken off early, some late. They're growing rye afterward. Crop variation is a huge thing in soil health.

It took me a long time to figure out that plants determine biology. This is why you have to rotate. The diseases and insects that digest corn don't digest alfalfa. Farmers who put on manure or who improve their crop rotation take a huge step forward in their success, because they have a diversity of feeds and a diversity of biology, and it eliminates a lot of their problems.

JOHN: This goes back to the discussion we were having earlier about biology being the most challenging thing to fix in the soil profile. There's this entire discussion in agriculture that says that it takes a healthy soil to grow a healthy plant. I think what we may have missed is that the fastest way to develop a healthy soil is to grow really healthy plants.

GARY: Yes. That's really interesting. Let's say you have a cover crop, with one plant per square yard—thick and lush and beautiful. You have to be willing to feed it to the soil at the stage of growth that affects the biology in the soil for the outcome you want. In other words, if I want to grow a crop like corn that recovers a lot of soluble nutrients, the crop I'm going to grow previous to that cover crop—what I'm feeding the soil biology—better be highly digestible and young and succulent.

It's like feeding a cow. If I want to get 100 pounds of milk, I want to feed a legume. I can feed a more mature, complex plant that would feed the fungal population better, and it wouldn't provide all the soluble nutrients. Then the plant itself would go to work, and the biology would work with it. That's the kind of understanding that has to take place within this system.

JOHN: Gary, what is something you wish all farmers knew?

GARY: I wish they could stand back and see that farming is a system that has all these little moving parts. Their job is to identify the limiting factors or constraints within their system and fix them.

Most farmers think, "Well, the bugs are in the air, and they're coming through, so I better spray." I was in England, and they developed something at Cambridge University that you put on the corn to measure fungal spores. If the fungal spore gets to a certain level, you spray. And I said, "Really?"

Plants can defend and protect themselves, and farmers don't understand that. They don't understand that trace minerals can do a lot to protect and make your plant healthier, and that nitrogen is a negative on fungal diseases. It causes trouble. They don't understand that they themselves have an impact on soil and plant health. They still have their conventional mindset—that every time there's a problem, you just spray it.

What in the world gave people the idea that we could buy a jug full of bugs, dump it on the land, and everything would be great? Why won't those bugs survive? In fact, why aren't they there now? The people selling them say the bugs extract natural organisms from healthy soils to put on an unhealthy soil. Why would you think that would work? Why wouldn't we just change what we do on the unhealthy soil?

JOHN: Exactly. And what makes you think that they're going to survive if you don't give them a food source?

GARY: They might see a bushel or two gain from doing this. But that's not fixing the system.

JOHN: GARY, the times I've learned the most are when something didn't work. When things don't work, you have to really dig in and try to figure out why they didn't work. What are some experiences you've had in which things didn't work the way you expected?

GARY: Probably expecting that if we just put on some calcium or some balanced fertility, that we would have less of a disease or insect problem. Let's say the nematode population is excessive, and we don't recognize or identify it. We're trying to do these wonderful things, but they're still out there. We haven't changed that.

I think we have to earn the right to not use an insecticide or a fungicide. When we had failures, maybe we were jumping the gun, and the farmer was anxious to be able to get rid of some of those costs. Then, of course, we had failures. Then it's hard to identify.

People said, "Oh, there's a nutritional deficiency out here."

I said, "Yeah, but someone just ate the roots off."

JOHN: Gary, you've written a couple of books. Do you have another book coming out? Are you working on a third one?

GARY: The third book I'm working on right now is about the Frank Lloyd Wright Taliesin Farm that we farm here in Spring Green, Wisconsin. We get a lot of visitors. We're writing a book about the history of agriculture—farming yesterday, today, and tomorrow—which gives me a chance to go through the history of agriculture in our community. It's an educational endeavor. The farm was under rampant chemical inputs for forty years and was run down. Nobody fixed anything.

That last book—what I just came out with a year ago—was a revised version of my first biological farm book. It's now a 500-page thing. I'm done with that book. I don't ever want to read it. I don't ever want to rewrite it again. That's enough! But I do want to do a dairy nutrition book.

I am a dairy nutritionist at a high-forage dairy farm. We took all the grain away from our dairy herd. Anything you add to the cows' diets was missing in the forages. Now granted—we probably won't get as much milk—but we ought to have healthy production. I'm trying to figure out what kind of forages and what kind of ratios it's going to take to make a high-producing or healthy-producing cow at the maximum level. When you take all the tools away, like grains, you're only going to have success if you're able to get more energy and more nutrition into the food you grow.

That book's down the road someday. That's the one that I want to write—about high-forage dairy farming.

JOHN: Obviously, the limitation that you have in removing grains from dairy is energy. How do you produce forage with high-enough energy to safely remove the grains?

GARY: There are certain plant species that will help us. But again, this is about getting enough active calcium. We've seen the soluble carbohydrates. We can grow a big, solid-stemmed alfalfa and get more pectins into that plant. We can raise that energy level. Really, that's why cows and soils are so much alike. That's why, as a dairy nutritionist, it was easy for me to do soils.

In cows, we have to balance energy to protein. When we give a dairy cow highly digestible feeds, we always struggle with excess protein. Then they're on the Atkins diet, which burns up their energy. We have to grow the right kinds of forages—those that balance energy with protein—and they have to be highly digestible. But we can have an impact on the amount of energy any specific plant takes in.

Our modern-day farms—these big megafarms—they feed an unbelievable amount of corn silage. They're putting a lot of energy into cows, and they can get a lot of milk out of them. Of course, the cows don't survive very long, and I question the quality of the milk and the fatty-acid makeup of it. But we can have a profitable, healthy herd of cows. That's our belief and opinion.

A lot of the 100 percent grass-fed guys don't get enough milk volume to pay their bills. It's easy to feed grain. It's not so easy when you can't use it anymore, when you have to get it all out of your feeds and can't violate the principles of that cow. The cow needs energy-to-protein balance. It needs certain minerals. How do I avoid excess minerals when I'm only feeding forages? Like you said, excess probably causes more damage than deficiencies do. That's true for soils and cows.

That's why it's a really big challenge to take that grain away. The grazing dairy herds out here that are all forages get about eight thousand pounds of milk. We had a 20,000-pound herd average when we took the grain away just six months ago. And yes, we've dropped in milk. It's a whole system, starting with genetics. We've been breeding for an overall smaller cow with a larger body capacity—a larger food intake capacity. The heifers that calved this fall, that were raised without grain, are totally different than the ones that we took the grain away from.

That's why I find it challenging and fun to learn. I hope we can get to a 12,000- to 15,000-pound herd average just on forages. That's our goal. If we can do that and have cows healthy and have high-quality milk, I think we'll please the consumers—those who are questioning the food supply. That's why grass-fed beef gets a lot of attention. But if that cow is unhealthy and sick eating grasses, that does not make a high-quality product.

JOHN: The other interesting piece is this tremendous dairy overproduction here in the country. If a substantial number of farmers lowered their production a little bit and tried to produce a healthier, higher-quality milk, that might fix the milk consumption problem as well.

GARY: Yes, I agree 100 percent. Right now, in a corn and bean world, growers are all trying to get more yields—maybe at an environmental cost—because they can't pay their bills. But then they flood the market, and then they really can't pay their bills. I think you're right.

We're dropping three, four, and five thousand pounds of milk on our cows. We've taken some of the surplus off the market. If a lot of people did that, we wouldn't have a surplus. My role is to make it more comfortable for people to change to an all-forage diet. That's what we're trying to achieve.

JOHN: Gary, in addition to the books that you have written, what is a book or resource that you would recommend to growers?

GARY: I think about that quite a bit. *Acres U.S.A.* is still a prime magazine for me. I probably get fifteen magazines a month. I actually enjoy reading strip-till and no-till magazines. There's interesting information that people are starting to bring up and talk about. You have to be able to put it in perspective, and you have to be able to read between the lines on that kind of stuff. For organic, I like the MOSES Conference we have here in Wisconsin. There are a lot of resources and people there.

I think you have to live your life being underopinionated instead of opinionated. I don't really care what your opinion is, and people shouldn't care what my opinion is. But they should recognize what your life experiences are. They should say, "How might that fit into my farming operation? How can I adopt some of those concepts and ideas?"

I was a really big student of Don Schriefer years ago. He taught me a lot about soil health and physics, and the strip-till system all came from him. I've read a lot of books, and there are a lot in the Acres library that would help people get a better understanding. Education is what gives you the comfort and ability to recognize that "Yes, this is a constraint on my farm," and "No, that's probably not."

JOHN: Indeed. Gary, what is the question that you wish I would have asked? What's the topic that you think is important to speak about?

GARY: I'm interested in the things you're doing with sap analyses and the products you're using and when you're using them. I've always thought that you need to understand the big picture before you get into the details.

I've spent my life and career trying to transition a lot of conventional thinkers into a more biological mindset, without giving them so much detail that they get bogged down, that they're not looking for some kind of magic, that they don't lose sight of the base.

JOHN: I think there's always the need to have more of those detailed conversations, as well. It's very important because there are many growers who, as you described earlier, are open to the system; they want to learn the system. Of course, learning and understanding the system are all about understanding the details that comprise that system.

GARY: When we first started out in this fertilizer business, the focus was on choosing better qualities. Instead of potassium chloride, we used potassium sulfate. Instead of DAP, we used MAP and rock phosphate. We chose better sources. Instead of using oxide trace minerals, we used sulfates.

Well, as time went on, we started manufacturing fertilizers and got involved in carbon-based fertilizers—trying to mimic what Mother Nature would do by taking minerals up in plants and having biology or food hooked to that. That's another whole discussion: How can we deliver minerals and sources through carbon? I think carbon, linked with fertilizer, is more efficient. I just read a report the other day where the researcher put a tracer on nitrogen and found that only about 25 percent of the nitrogen farmers apply is sucked up into the plant. The rest of it is either lost or is going into the biology or is hooked into some carbon somewhere along the way.

JOHN: It doesn't surprise me at all. When we started using sap analysis on farms to see what plants were actually absorbing, on many farms—I would say on most—we can cut nitrogen application rates from 30 to 50 percent or more, with no decrease in the nitrogen levels in the plant.

GARY: If you truly want to sequester carbon, you have to back off of nitrogen because you're burning carbon up with nitrogen. If you don't back down on nitrogen, you're never going to build carbon. How can a farm be conventional all those years and put all those corn stalks back in the ground and keep knifing-in anhydrous? This organic matter is at best stabilized, and it hasn't dropped. Residues didn't build it up. Nitrogen burnt it up. And they were no-till.

JOHN: It's interesting how you mentioned Don Schriefer. One of the things that Don Schriefer wrote about in his books *Agriculture in Transition* and *From the Soil Up* was that the fastest way to build soil organic matter is to grow corn. Today we think that growing corn depletes soil organic matter. It's not about the crop. It's about the way we manage the nutrition, particularly the nitrogen within that crop.

GARY: That's right. That's why I'd tax nitrogen. The university says 1.1 pounds of nitrogen per bushel of corn. We got guys operating at 0.7, 0.5. They're still corn and bean farmers, but they're much more efficient, and their mineral levels and their biology are up.

Insects: Nature's Garbage Collectors

TOM DYKSTRA

The more a plant photosynthesizes, the higher the Brix reading. The higher the Brix reading, the healthier the plant. And when you have a healthy plant, you don't have to use, for example, all of the pesticides that are being used today: herbicides, insecticides, fungicides, nematicides, etc.

You can really prevent soil erosion from occurring by having good healthy soil, whereas if it's not cohesive and it's not bonded together through the glomalin or some other thing, yes, you can lose soil quickly.

Once you get above twelve Brix, insects really aren't causing any more issues with your plants. And once you get above fourteen, they're really not even landing on your plant, unless they just want to rest on it, because they won't be able to take a bite. Or if they do take a bite, they won't be able to get through the cuticle or into the phloem tissue; it's not going to be digestible to them.

Insects are only tuned in to the unhealthy plant. No insect will ever attack a healthy plant. What they're zooming in on is the unhealthy plant because it's digestible. Healthy plants are not digestible.

Nitrogenous fertilizers are salt-based. It's a force-feeding operation; the plant is not really feeding itself. The salt-based fertilizer increases the conductivity and forces nutrients into the plant, causing a quick burst of growth. But all along, the plant is stressed, and the insects know that.

IT'S DEBATABLE whether modern farmers are more concerned with killing weeds, killing diseases, or killing insects. Any one involves the massive application of synthetic compounds—substances that have adverse effects on soil and plant health. But what if insect problems could be mostly eliminated simply by improving the health of the plants we grow?

Dr. Tom Dykstra believes that this is possible, that insects only attack unhealthy plants. By improving the health of our plants, we can reduce or eliminate the need for caustic chemical pesticides.

Tom is an entomologist—one who studies insects—and he earned his PhD in bioelectromagnetics at the University of Florida under the tutelage of Phil Callahan, the great pioneer of how plants and insects interact biophysically. Tom is the director of Dykstra Laboratories in Gainesville, Florida. The lab studies such subjects as insect olfaction (how insects smell), insect carbon dioxide detection, and chemoreceptive proteins.

Tom has done some extraordinary work in the area of entomology and plant-insect communication systems. His insights are practical and are sure to benefit all types of growers.

JOHN: Tom, you've done some fascinating research over the years, both on your own and with your mentor, Phil Callahan. What are some of your most memorable moments that led you to where you are today?

TOM: One of the most memorable moments for me was probably getting fired at the University of Florida, which led me to study under Phil Callahan. That would probably rank near the top! Had that not happened, I certainly would not have been on the path that I am right now.

I was working at the time with flea larvae. It was a seemingly uninteresting insect—or so I thought at the time—that led into my PhD project and allowed me to explore the world of insect bioelectromagnetics, which then opened up a whole new world to me that I had never been exposed to—either in academia or in the literature and even less in books.

This has been an exciting journey for me. The reason why is because it's a new field, and the only people I know who are even remotely in the field are in the military. Within academia, it's quiet on this front. But I do enjoy what I'm doing here, and I'm happy that I've had a chance to explore this fascinating area of insect bioelectromagnetics.

JOHN: Insect bioelectromagnetics sounds fascinating. Can you describe for us what that is? What are you actually working on?

TOM: The word has to do with electromagnetic waves and how they are influenced by organisms. Thus, the name "bio." So electromagnetics is simply the pure waves, and bioelectromagnetics is how these waves influence life.

These waves could be external to the organism—environmental—or they could be internal: the communications going on within the organism as to what information needs to pass from one organ to another or one part of the body to another or, in some cases, to actually amplify the message to make it more easily discernible to an insect or even another animal.

The field of bioelectromagnetics is relatively large. There's a journal, and there are many journal articles you can find.[6] But the specific field of insect bioelectromagnetics is not. It's a much, much smaller field. Those who are involved in the insect side of it are few and far between. The field itself is relatively small, which I guess gives me a certain amount of anonymity, which is required because I do run a research laboratory, and we are not allowed to publish. Most of the research we do is kept under wraps. We've had some wonderful discoveries that I really hope I'll be able to divulge to the public someday.

JOHN: When you speak of insect bioelectromagnetics, what is the impact or the potential impact of your work on agriculture, and what might farmers be able to apply in the future?

TOM: That's a difficult question because when you get involved in pure bioelectromagnetics, you're looking at fields that really are not measurable and, for the most part, should not be measurable because the farmer is not going to be out there with the type of equipment one uses to measure these fields. In a direct sense, there would be very little impact. In the more general sense, there is a much larger impact, and it's not something that most people are aware of: energy and how it is acquired by plants.

---

6 "Bioelectromagnetics Journal," Bioelectromagnetics Society, https://www.bems.org/journal.

Photosynthesis is certainly a well-known phenomenon that everyone is familiar with. It's the starting point in the field of bioelectromagnetics, just as it is in agriculture. If you can optimize photosynthesis—which has an electromagnetic component, with biochemicals and organic chemistry—you then optimize Brix levels because the relationship between photosynthesis and Brix is about as linear as you can get. The more a plant photosynthesizes, the higher the Brix reading. The higher the Brix reading, the healthier the plant. And when you have a healthy plant, you don't have to use, for example, all of the pesticides that are being used today: herbicides, insecticides, fungicides, nematicides, etc. It doesn't matter what -cide you're using; they're all doing the job of killing life. As a farmer, you're trying to augment life.

This augmentation can take many different forms. The sun and the energy that comes from it is certainly an obvious one. There are less obvious electromagnetic waves out there. The Schumann frequencies are relatively well known and have been studied by atmospheric sciences for a long time; they have their influence too. Once you are surrounded by all these frequencies, to be able to absorb the energy or to use them to communicate, then you need to have healthy plants, healthy crops, and a very healthy environment. Understanding that and ways to optimize the electromagnetic component is only going to benefit farmers.

If I were giving a talk, I would want to help farmers understand the fields and how to optimize them and how to prevent the farmer from getting in the way or putting up any blockages to these frequencies.

JOHN: I would love to hear more of your thoughts regarding photosynthesis. Photosynthesis is so foundational. When you're able to increase it, you have all these other benefits—increasing carbon sequestration, improving plant health, disease and insect resistance, etc. This is something that we focus a lot of attention on. We often approach it from a biochemistry perspective of thinking, "How we can optimize photosynthesis?"

How much do you think it's possible to increase photosynthesis above what is common or normal, depending on how you chose to define that word? How much can it be increased? Looking at photosynthesis from a bioelectromagnetics perspective, what tools could we use to optimize it and to increase it, if possible?

TOM: You can increase it substantially. The reason I know this is because most of the crops that I've had a chance to look at are between four and eight Brix. Once you get up to roughly twelve, all serious insect damage stops. And once you get up to about fourteen, you have a genuinely healthy plant that is producing healthy food that can be consumed by a human being and will lead to a long life.

But because we are so far below what is optimal—optimal would be fourteen and above, and twelve is the minimum so that you don't get insect damage—I can tell you unequivocally that we can increase photosynthesis considerably, simply because we have so many plants right now that are suffering—suffering in the sense that they're not working up to their potential.

Even getting to nine Brix, which is still not a healthy plant, is an improvement over a lot of the plants I've had a chance to look at. Whether plants out in someone's backyard or crop plants on the farmer's field, they're all dealing with very low Brix, and therefore, the plants are compromised. The only way we can go is up because once you go down, plants die, and there's really nothing that can save them at that point.

If you want to maintain a plant at a very low Brix—let's say one or two—you have to feed it every day. There's no way around it. Anyone who has ever played golf has walked on plants that are running about one or two. If you go even one day without feeding that grass under your feet, it will suffer considerably because it has very little photosynthetic part above the ground. Its root system has been absolutely compromised, usually due to what has been applied. And so that's the extreme case, where you can actually—in a very artificial way—keep a plant going by continually feeding it. It's a force-feeding operation.

The same thing goes with crop plants. Now with the crop plants, the farmer is not out there feeding them on a daily basis like you would find on a golf course. But the farmer is going to be feeding his crops. At a minimum, he's going to be doing it once or twice during the season but probably more often than that, depending upon what type of plant it is, whether it's a perennial or not.

It's a matter of getting the photosynthetic rate up. As I alluded to, this means not putting up any of the blockages. Some of these blockages are certainly and absolutely pesticides. Pesticides are killing all of the microbes in the soil. And the microbes can be selectively killed. So one herbicide is going to selectively kill certain microbes. An insecticide is going to kill a certain set of microbes as well. A fungicide will kill a certain set of microbes. Even a nematicide. We have nematicides down here in Florida. They're not as common in the rest of the country, but each one of these is doing its job to kill.

When these microbes are gone, they are no longer able to hold on to the energy that is in their bodies. You have now given up a massive energy source. This is where the electromagnetics comes in: if you do not have those microbes, there is no energy in the soil. Each microbe is about half a volt of sequestered energy. This doesn't mean that there is this massive amount of energy—that if you stepped on some ground with millions of microbes, it would blow your foot off or catapult you into the air fifteen feet. What this means is that this is clean, stored energy that's found in the membranes. It's found in the chloroplast, in the endoplasmic reticulum.

Everything that is inside the cell has a nice, clean storage form of this energy, and that energy is then made available to the plant, either directly or indirectly. Sometimes the plant can use the nutrients if the microbe dies. Microbes are dying all the time, and some of their nutrients can be taken up by the plant because everything is bioavailable when a microbe dies. Or you can have a more indirect result—when the microbe is actually releasing minerals to the plants. These microbes can break down minerals far more efficiently than plants. Without these microbes, the plant is no longer able to take in minerals. These minerals are the constituents—the cofactors—that are used to keep photosynthesis going. Without these minerals, photosynthesis suffers, and we see lower Brix.

Everything seems to be connected together through the microbes—the energy—because plants want to put their roots into a high-energy soil. If it's been sterilized, they don't like that—it then becomes incumbent upon humans to force-feed the plant. And this is not the way plants were designed to live—to be force-fed on a regular basis.

This is the importance of energy. It's not just about that which is above the ground—the leaves and photosynthesizing. It's also about getting those minerals from the soil so that you can assist photosynthesis going on in the leaves. These are all important and have a component in making sure that the energy systems of plants are revved up and doing exactly what they're supposed to be doing.

We can easily measure this with a very simple measurement of Brix. There are many different ways of studying plant health. Brix is by far the easiest thing to do if you're a farmer. There are other, more complex ways. You can send things out to a laboratory. You can get tissue analysis. You can play all types of complex games, and scientific laboratories can get involved. But if you need something that can be done right there on the spot, it will be Brix. It's an easy way farmers can tell whether their plant is healthy, and it will have an impact on their crop—and of course then with their yield and money in the long run.

JOHN: One of the tools that we've been using a lot in lieu of the refractometer or Brix meter is sap analysis because it actually measures sugar content. The refractometer is a very useful and valuable tool, but the challenge that many growers have with it is its variability, and the inconsistency that can occur based on moisture and sunlight levels. It's something that we do use, but we actually use a lot of sap analysis results as well to try to remove some of the variability.

You described how you're essentially looking at plant growth from a biophysics perspective rather than from a biochemistry perspective. And you're saying that energy for plants doesn't just come from photosynthesis and sunlight. It actually comes from soil biology because of the voltage contained within the microbial cells in the soil profile. And that if you really want to increase that energy, the first step is to do no harm and to stop killing the biology. Is that a summary of what you're describing?

TOM: Yes, that would be one of the first steps—to stop using pesticides if we can help it. There are some individuals who will need to wean themselves off over the course of one, two, or maybe even three seasons. But to be honest with you, I'm not sure if I would be an advocate of three seasons. I think that weaning yourself off should be done as soon as possible—certainly within a season if you need to. There are ways to speed the process up—to get the soil healthy again—so that you can start farming the way that God intended.

JOHN: What are some of those ways of speeding up the process? I think that's the question that I'm really trying to get to. Obviously, we can discontinue use of these pesticides. But then what can we do to increase the energy flow in the soil profile?

TOM: What I usually ask a farmer with, say, one thousand acres is, "How much are you willing to give up? How much can you afford not to produce on?"

Sometimes they come back with "I can't afford anything."

But when you really press them, they'll tell you something like five to ten acres. That would be an excellent starting point to work from.

I would recommend that they allow the ground to go fallow for at least one season. What you're doing is allowing the weeds to grow that want to grow on that soil because they're going to do their best to improve it. You want to disc the weeds into the ground because all of that organic manner needs to go back into the soil. Make sure it's not on the surface because it will get burned off in a slow oxidative process by the sun. Every time the weeds get to be what you consider to be bad, go out there and disk them back in. The second thing you would want to do is to apply about five pounds per acre of sugar as often as you can afford. That's different for every farmer.

Why would you want to do something like that? When you put sugar out, you increase the microbial content tremendously. The fastest way to get rid of toxins in the soil is through microbes. Why? Because the microbes have the glutathione S-transferase. They have the esterase. They have the P450. They have all of the equipment necessary in order to break down those toxins. You want to increase the level of the microbial activity tremendously—almost to an unhealthy degree—and get the microbes cleaning up the soil. They do not like to be in with all those pesticides, and they will do their part to clean it out. The fastest way to do that is to throw massive amounts of sugar at the ground. There are some other things that can be done. They're sometimes more expensive, and if you have deep pockets, you can go that route. But this is usually the simplest and the cheapest way to clean out your soil in as little as one season. You can do it in less than one season, depending upon how toxic the ground is.

Once you've started to clean things out, and you're discing weeds back in, you can improve the ground tremendously in about one year. You're going to be able to plant on that. It's not like I'm destroying the farmer's livelihood. He's still going to be able to take care of his family. If he's got almost a thousand acres left, but his production on that five acres will far exceed what he's getting on the rest. This is a process that needs to continue; it's done rather painlessly.

There are some individuals who would like to switch over immediately. That's a little more painful. But if you want to do this a little bit slower, you can do so relatively inexpensively. That is the best way to clean things out. If you continue to use the pesticides because you feel you have to, then I would choose some high ground to start with the five acres. The reason why is the runoff—I don't want any pesticides getting on that land that you've just spent time cleaning up.

Now, if you let the ground go fallow for ten years, in almost 100 percent of the cases, everything will be taken care of. Studies have been done on this, and it's relatively well documented. But no farmer wants to let his ground sit for ten years.

The fastest way is not through chemicals. There's no chemical you can get from Sigma Chemical or any other place that's going to get rid of and detoxify those chemicals better than microbes. By putting the sugar out, you increase the microbial content so high that you probably wouldn't even want to be putting plants on there.

But that's not the idea. The idea is to clean your soil up as quickly as possible. If you can put five pounds down a week, that would be ideal. You will clean up the ground very quickly. Some farmers have tried to put on ten pounds. I don't usually advocate that because I don't want it getting sticky, and I don't want the nozzles getting all plugged up. But some people have compensated with a little bit more water. That's your choice. If you're putting down about five pounds per acre, you're going to be in good shape. That will be the easiest and the fastest way of not only increasing the carbon content of the soil but of getting rid of those toxins. You will have saved yourself many, many years of trouble. This is a process that would have happened over time, but you can speed it up, considering that sugar goes for about one dollar a pound. We're not talking about a lot of money here. If you can afford five dollars a week, then all you have is the cost of going back and forth over your field.

This does go along the lines of cover crops. That's very similar to what I am suggesting. It's allowing things to grow. Sure, you can plant some cover crops, and they can do their job too. That's fantastic. But the thing with cover crops is that there are many different ones out there. Some people like to do one. Some people have compaction problems, so they want to use radishes. Down here in Florida, we use ryegrass a lot because we have compaction problems. But sometimes there's the difficulty of choosing what cover crop you want. Cover crops will do a very quick job of removing toxins from the soil, but not quite as fast as the sugar. When you put out pure sugar, you will see a massive spike in the microbial content that you would never get from a plant or from a cover crop.

Plants release between 20 and 70 percent of their photosynthates back into the soil, depending upon how healthy they are. It's a slow process. It goes on twenty-four hours a day, especially at night. They're introducing the sugar into the soil—but they may have some difficulty doing it as quickly as just giving the soil a massive dose. Once the cover crop has taken hold, you can back off of the sugar. You're going to be in pretty good shape at that point.

But make sure that the cover crop is disced back into the soil. The mistake is to essentially lawn-mow it, with everything laid out on the surface—because they're thinking about soil erosion and things of that sort. But soil erosion will be taken care of. You can have bare soil and have very little soil erosion if you have a high microbial content because a high microbial content leads to a high fungal content, and a high fungal content produces glomalin. Glomalin will stick things together. You can really prevent soil erosion from occurring by having good healthy soil, whereas if it's not cohesive and it's not bonded together through the glomalin or some other thing, yes, you can lose soil quickly. But honestly, erosion is not as much of a problem as in decades past. It has only recently become a huge issue because we have sterile soil. Nothing is holding it together. Anytime wind or water comes along, it is easily picked up.

JOHN: Tom, I think this is a fascinating conversation because many agronomists agree that it is very important to keep soil covered with green, growing plants—preferably a diversity of plants for an extended period—as much of the year as possible. When you look at the mathematics of what you just described, if most plants are contributing 20 to 70 percent of their photosynthates to the soil profile, isn't that a much larger quantity than five pounds of sugar per week? Why would cover crops not have a much bigger impact than the sugar itself?

TOM: The cover crops take time to grow. Most of the individuals who are applying sugar have ground where nothing is growing. Over time, the cover crops will catch up. And you are correct. When you have a healthy amount of cover crops, they're going to exude more sugar, assuming their Brix levels are high enough. I should have prefaced this with the fact that some of the farmers I speak with are dealing with dead soil. They have very high electrical conductivity, high-salt soils. Nothing is growing, not even foxtail and kochia. They're in a situation where they need to revitalize the soil, and their salt problem—as well as their pesticide problem—can quickly be taken care of because they've got bare ground. They can do this by throwing sugar.

Sugar is usually a great way to start. If you are doing a cover crop soon after, it should be able to catch up relatively quickly, depending upon the cover crop. A variety is better than using one cover crop. There are some advantages to doing just one, and this has to do with machines: it's easier to go over something if it's the same cover crop rather than having a variety. But if you've got the capabilities of taking pretty much any of the cover crops and throwing them back into the soil, whether through discing or some other means, then not only will you have the sugar being thrown into the plant by the cover crops; you will also have the organic matter, the carbon under the soil, which will be broken down by the microbes. That will then add considerably to the carbon sequestration of the soil.

Excellent point! You're right. There are various situations that would call for cover crops instead of sugar.

JOHN: Thanks for clarifying that. You mentioned that erosion has become a much bigger challenge recently than it was historically. Why do you believe that is the case?

TOM: Because the microbes are gone. Since the microbes are gone, you have fluffy soil that is easily washed away. Whereas when the microbes are in there, they're dying all the time. And when microbes die, their membrane breaks and liquid goo comes out of their system. There are millions upon millions upon millions of microbes dying every second in your soil. And that, in and of itself, is a type of goo that helps keep the soil together.

Imagine putting soil under a heat lamp and sterilizing it so that you've got nothing in there right now. It is very easy to wash that away. You're going to have no moisture. You're going to have no microbes. You can blow on that type of soil or sand, and it will easily get blown away. Or you can put some water on it and it will easily get washed away. This is a situation that was not as serious in decades past but has become more serious with the advent of the pesticides, which have helped to sterilize the soil and have made it far easier for soil to be picked up and deposited somewhere else.

JOHN: What is something that you puzzled over for a long time?

TOM: We have puzzled for over twenty years over how insects smell. This is something that has been a big issue for me.

We knew that they were smelling electromagnetically, but we did not know the process and how it was occurring. When I say, "puzzled over," this is probably at least ten years—pushing fifteen—when I was unable to put the pieces of the puzzle together. Even though we knew it was electromagnetic, the process was not well worked out.

Understanding how photosynthesis and rhodopsin and some other proteins work, about absorbing energy—how ultraviolet energy is absorbed by certain amino acids—and understanding how this process occurs—this has helped me to understand how insects are actually smelling. As of November of 2016, we were able to decipher the insect olfactory code. We were finally able to understand what frequencies were involved.

Even though we finally—after practically my entire professional life—figured out how they were doing that, now that we've got the system down, it's a matter of working out the details: understanding not just how insects smell carbon dioxide but to explain why carbon dioxide is detected differently among *Culex* mosquitoes, *Anopheles* mosquitoes, *Aedes* mosquitoes, and so forth. Why bedbugs smell carbon dioxide differently, why kissing bugs smell it differently, to understand why some *Lepidoptera* butterflies and moths do so differently. We're getting into the details right now. Yes, we understand how they're doing it, but now, we understand that each insect is smelling differently.

JOHN: This is a fascinating piece, Tom. I know that this is something that many growers are very interested in and want to understand much better. Would you be willing and able to dig a little bit deeper into describing how the olfactory senses work and why some insects are attracted to certain regions where others are not?

TOM: Generally, the insects are smelling with their antenna and with their palps, which are some of the mouth parts. When they're flying through the air, they pick up various odorants in the air. These odorants are vibrating and giving off energy. They're absorbing and emitting energy constantly. As long as they are above zero degrees Kelvin, which is absolute zero, they are vibrating.

These electromagnetic vibrations can be easily detected by various equipment in the laboratory, as well as the equipment on an insect. They have various sensilla. I would call them tiny antenna on the actual antenna-proper of the insect. It's these antenna—whether it be the antenna-proper or whether it be the smaller sensilla on the insect—that are tuned into very particular frequencies. These particular frequencies are all important for the particular insect.

Some insects would not be tuned into CO_2. They would have no reason to be. Whereas other insects, like mosquitoes, would be tuned into CO_2. There are certain floral compounds given off by plants that some insects are going to be attracted to—honeybees being prominent among them. Then you have certain scents that are not attractive at all because some insects don't go after flowers. There are some plants that advertise themselves as unhealthy. And when they do so, they are picked up by other insects.

Insects are only tuned in to the unhealthy plant. No insect will ever attack a healthy plant. What they're zooming in on is the unhealthy plant because it's digestible. Healthy plants are not digestible. Unhealthy plants are. Because they can't digest a healthy plant, there's no interest in even attacking it. It either has to be injured or it has to be unhealthy. By doing so, it is now digestible, and this is what an insect is going to attack.

At the beginning of our conversation, I mentioned how once you get above twelve Brix, insects really aren't causing any more issues with your plants. Once you get above fourteen, they're really not even landing on your plant, unless they just want to rest on it, because they won't be able to take a bite. Or if they do take a bite, they won't be able to get through the cuticle or into the phloem tissue; it's not going to be digestible to them. They'll have to pull out and move on to a different source.

All insects are flying above, looking for crop plants that are digestible to them. Plants advertise themselves as being unhealthy, essentially saying, "I'm unhealthy. Please come eat me." And so the insect will come in, and it will start eating the plant because that's their job.

Our job is to eat healthy plants. We have a much more elaborate enzyme system. Our digestive systems are designed to eat healthy food. We don't do well eating Doritos all the time. We do well when we eat healthy plants, and this leads us to be healthy.

For the insect, it's different. Insects do not do well on healthy plants. They can starve. You can take a Colorado potato beetle and put it on a healthy potato plant, and it will not be able to eat it. But you can take an unhealthy potato plant and the Colorado potato beetle will go to town because this is what it is meant to do. Because insects are only attracted to unhealthy plants, the unhealthy plants will advertise themselves. They've got certain visible frequencies that insects can detect, especially from a distance. They've got certain odors, and these odors can be picked up by the antenna and by the sensilla, and the insect will move in for the kill, so to speak.

JOHN: When we look at plant health, we're talking about healthy plants versus unhealthy plants. What are some of the compounds that serve as insect attractants that we could manage and monitor?

TOM: You can't. It would really have to be a general thing. Generally speaking, ethanol is a universal odorant that advertises itself as being unhealthy. So a lot of the plants will release not just ethanol but also various alcohol components. Not all alcohols, but many alcohols, advertise a plant as being unhealthy. It's a hallmark of fermentation. Fermentation produces the alcohol.

When a plant is degrading and it's in trouble and it's fermenting, even in a small way—even in an imperceptible way—it will advertise itself. If these odorants are being released, they will be picked up by insects. It will change how the plants are perceived. You can take satellite images of two crop plants, and they look different on various images. It can be a visible image. It can be an infrared image. But they both may be corn. They both may be soybean. They both may be anything you could think of, but they will not have the same look under an infrared camera or under a visible camera.

This is something which is very profound in grasshoppers. You don't find them so much in the United States, but on other continents, locust swarms are a problem. These locust swarms are not just millions of individuals but billions—sometimes trillions—of insects. They descend upon a very particular crop and take it all the way down to the roots and then pick up and fly away. They will leave a farmer's field right next to that exempt. These are the remarkable things that you realize when you see stuff like this. The grasshoppers made a decision. They made a decision to eat one plant over another. Why? Why didn't they just come down and eat everything? We've always been told that grasshoppers will eat anything, and yet there is direct proof in some of the images that I have seen and testimonies of others that, no, they actually are very selective.

I should tell you that grasshoppers are less selective than other insects. Some insects will disappear by a Brix of eight. Other insects will continue to chew on your plants right up through ten, eleven, or twelve Brix. But once they get to about twelve, they will lose interest. And the grasshoppers are among them. You can find the grasshoppers among slightly healthier plants for that reason, but you're going to find that with the aphids, the leafhoppers, some of the other Hemiptera insects, once the plant gets to eight, they lose interest in the plant. You just won't find Hemiptera insects on a plant above eight Brix. Those are the ones that have the beak that they stick into the phloem tissue and take a sip from the sugar water that is flowing around the phloem tissue.

Every insect has its own cutoff. You really have a lot of insects fall off by the time you get to eight. And as I mentioned at the beginning of this, most of the crops are between four and eight. A lot of plants are really susceptible to every single insect that is out there. But if you can get above eight, you pretty much can take care of your aphids and your leafhoppers and psyllids. The Asian citrus psyllid is down here in Florida, and other psyllids, because it's very rare to find a citrus tree that's above eight. We've tested a lot of them.

Yes, sometimes it's a little difficult to predict Brix because there are conditions that allow it. They're all sending messages. As a scientist, I'd like to decipher these codes to try and understand what the plant is telling us when it was at a Brix of ten one day and the next day went down to three. There are very specific reasons as to why a plant would do something like that.

Down here in Florida, we have storms all the time. Plants—in a very gentle, freak-out sort of fashion—when a big storm comes in, they shunt all of their sugar down into their roots. You could have drastic drops in leaf Brix levels. We have tested plants that looked healthy to us, and we'll test them right before a storm, and they'll come in at one to three Brix. And you think to yourself, "How is this plant even surviving?" Well, it's a temporary thing. When a major storm comes in, if the top of the plant is lost, the sugar that's in the roots will allow the plant to regrow once the danger has passed. Those leaves in the one to three Brix range, when we pull out some of their roots, they're measuring fourteen Brix. Not because the roots are normally at fourteen Brix but simply because it's a temporary shunting of the sugar around the plant. These are little clues that help you to understand Brix a little bit better.

People will come up to me and say, "Well, I found some psyllids on my citrus plant," or "There are some leafhoppers here. And we tested it, and the plants were coming in at nine. I don't see what the problem is." Well, the plant is never at a uniform level of nine Brix. There are certain regions of the plant that are going to be a little bit lower than others, and those are the ones that will get attacked. In the case of citrus greening, for example, I call it a "plug disease." A "plug" meaning it forms a carbohydrate plug that prevents nutrients from passing into a particular branch, and this causes the branch to go down in Brix. Therefore, that branch will become susceptible. It will advertise itself. The Asian citrus psyllid will come in. Because of this, that branch will be affected, but the tree may be a little bit healthier due to the plug, because things are not flowing around the plant the way they should be.

JOHN: We use sap analysis with many of the farmers that we are working with, and one of the fractions that is measured is the nitrogen profile. We're looking at the total nitrogen in the plant and also the nitrate nitrogen and the ammonium nitrogen. We've observed through field experience a correlation between elevated concentrations of, say, nitrate or ammonium and specific groups of insects. Are those compounds serving as insect attractants?

TOM: I would say, indirectly, yes. The reason why is because when you say "attractant," usually you smell something and then you go to it. That's how attractants work. Whereas the high nitrogen that you're referring to is not what I would consider to be an attractant. Nitrogen molecules are not emanating from the plant and being picked up. So it would not qualify as an attractant.

But what is going on is that you have an imbalance in the plant. It's this physiological imbalance in the plant that causes the nitrogen to be shunted in between one form and another. It can be a healthy form or an unhealthy form or a reproductive form or a vegetative form. That's the way the plant is now advertising itself as unhealthy, and it can do so almost directly in the electromagnetic spectrum. There will be visible frequencies or infrared frequencies that will be picked up immediately by insects because the plant is stressed.

For example, if you have a plant that's suffering, you look at it and say, "Okay, this plant needs a little bit of help." Let's say you put down a nitrogenous fertilizer—a classic NPK that you get in most stores. So we have a quick bloom. The plant greens up. Everything looks good. It looks like the plant has just gotten healthy. And to a certain extent, it has, but what you've done is you've stressed the plant because nitrogenous fertilizers are salt-based. It's a force-feeding operation; the plant is not really feeding itself. The salt-based fertilizer increases the conductivity and forces nutrients into the plant, causing a quick burst of growth. But all along, the plant is stressed, and the insects know that.

Imagine you have a grass plant—a monocot. If it starts to experience drought, its water levels are going to go down, and the plant will start to suffer. Many individuals would say, "I can see that the plant is in trouble right now. I need to feed it." They're not necessarily thinking it probably just needs water because of the drought; they think it needs something else. So you put down a nitrogenous fertilizer in order to green things up. And this is very, very attractive to the chinch bug—*Blissus leucopterus*: (1) there are drought conditions; (2) there's high nitrogen. When you have that combination, you're going to bring in the chinch bug because this is what it likes. You will notice down here in Florida, and elsewhere, that the chinch bug does not attack grass indiscriminately. It has to attack grass under very specific circumstances. It's not just about the drought, nor is it just about putting down the nitrogenous fertilizer and throwing excess nitrogen into the plant. The insect is looking for a very specific combination. The type of stressed plant that it's looking to eat is going to be more easily digestible under certain circumstances.

Even though one million insects have not been deciphered, we understand how some insects work and why some insects are attracted to some unhealthy plants and other insects to other unhealthy plants. Citrus down here, for example. We have multiple insects that attack citrus, but you usually don't find all of them feeding on exactly the same plant and exactly the same population. You'll find the Asian citrus psyllid focused on one area, the leaf notchers located in another, and the citrus leaf miner on another. These are all clues that, yes, the plant is unhealthy. That's a universal constant. But the conditions are different for each insect because each insect has its role in eating that which is digestible to it. The citrus leaf miner is inside the leaf. The Asian citrus psyllid is going after the phloem tissue. The leaf notchers are going after the roots. They'll go ahead and take a bite of the leaf. But they've all got different parts of the plant, and the plant has to be of a certain state of unhealthiness in order to be attacked.

This is why the generalities of Brix have been lost on a lot of people. It really is more complex than we think. I know that in the past, everyone said, "Get to a Brix of twelve, and the insects will stop." That's not absolutely true. I have seen insects take a bite of a plant above twelve Brix. They will not stay in the seed, but they will take a bite from it. This looks like a contradiction, but it's really not. To understand what really is going on with Brix and using that as an idea to understand what the insects are doing allows you to make these connections. This should help the farmer. I can go out, take a look at what insect is attacking a plant, and give you a pretty good idea of what the Brix is going to be before I even test it.

JOHN: Earlier, you mentioned some of the compounds that plants are releasing, and also the signals that these compounds are transmitting at various distances. In general, what are the possible detection distances for some of these compounds that serve as insect signaling mechanisms?

TOM: The same as any other odorant. It actually doesn't really differ at all. It doesn't matter whether it's ten yards away or whether it's a mile away. Any of these odorants can be detected from great distances. As long as an odorant is able to be blown away by the wind and an insect can smell it, the insect will follow the scent of that odorant to its source. The distances can be tremendous.

As far as the electromagnetic, that really depends upon the electromagnetic components. That would depend on the compound eye. How far can the insect see? Grasshoppers are flying pretty high up in the atmosphere, so when they're looking down, they're not looking at individual plants like a lot of insects would. They see what we see when we look at satellite images. They see a bunch of squares, rectangles, and trapezoids. When they see a trapezoid that looks attractive, they will descend upon that trapezoid. Those are the plants that are sending off a general electromagnetic visual type of message; the odorants are generally not getting up into the atmosphere when a locust swarm passes by unless they're flying low.

Other insects will key in on a very specific plant. Sometimes they will even key in on the part of the plant that is suffering. Like I said, a branch that may be dying is where the insects will congregate. The distance is not really that important. They're going to be able to detect it no matter where they are, just as they would any other odorant.

JOHN: I have friends who will attract every mosquito for dozens of yards around, and others who, for all practical purposes, are mosquito immune. Can you describe the differences?

TOM: I hate saying this, but that is the difference between healthy people and unhealthy people. If you have digestible blood, the insect can use your blood. If you have healthy blood, the insect is not attracted to it. Does this mean you're healthy all the time? No. People have certain states of healthiness and unhealthiness. But as long as your blood is digestible, it's going to be attractive to an insect.

It's the digestive system. Insects do not have very good digestive systems. Stuff has to come in digested to them because they cannot digest things. They just don't have the enzymes. They're garbage collectors. They eat muck. They eat garbage. They eat bad stuff. They eat stuff that we don't want to eat. Take a look at cockroaches, for example, or fruit flies. They're always around decaying fruit. They're not around healthy fruit. These insects that are keying in on very specific plants. Or even, let's say, the fruiting structures of a tomato—that's what they are keying in on.

Mosquitoes need something that's digestible. When they take a blood meal, it must be incorporated through the process of vitellogenesis: taking those nutrients in your blood and incorporating them into an egg. This occurs in about twenty-four hours—very, very quickly. If they're not getting digestible blood, they cannot, through the process of vitellogenesis, take those nutrients and feed their eggs. That's a problem, because obviously the mosquito wants to feed its egg. The only reason it takes a blood meal is to feed its eggs.

We do have these differences among individuals. Some are just more digestible than others, and they're going to be more attractive. Grasshoppers have a choice as to what crop they eat, and mosquitoes have a choice as to what blood meal they wish to get. Through their antenna and all of their senses, they detect that someone is more digestible than another, and they're going to go after that individual. Even though that may offend some people, I do feel an obligation to tell the truth the way that I know it, according to the information that has been given to me for the past twenty to thirty years of my life studying insects.

JOHN: I very much doubt that you're offending people. I would say that you're simply communicating some very intriguing ideas that many of us have observed to be true in real life and have wondered about the differences. So thank you for sharing that.

TOM: I'm happy to do so. When you go to the supermarket and you see fruit flies flying around the tomatoes, some people think, "Oh, I don't want these tomatoes. They have fruit flies in them." But honestly, all you have to do is find that one tomato near the bottom that's injured, and that's where all the fruit flies are located—near the one that's unhealthy. That essentially equates to that which is digestible. A tomato that is degrading—that is decomposing—is being broken down. When it's being broken down, it's digestible. Insects will only eat that which is digestible to their system.

JOHN: You and I have discussed before how environment determines genetic expression. And you expressed that you have had difficulties infecting some mosquitoes with malaria. As we've been having this conversation about mosquitoes, that popped back into my mind. What are the differences between healthy and unhealthy mosquitoes as vectors of infectious disease?

TOM: Well, I personally have not infected mosquitoes with malaria, but I've talked to researchers who've worked on this. Let me put it this way. Most *Anopheles* mosquitoes do not transmit malaria. Most *Aedes* mosquitoes do not transmit dengue, yellow fever, or Zika. Most *Culex* mosquitoes do not transmit encephalitis.

When you observe this, you understand that not all insects are infected with the disease. You can get bitten by thousands and thousands of *Anopheles* mosquitoes and not get malaria. But all you have to do is be bitten by one mosquito that does have malaria in order for you to get it.

This is one of the truths of biology that has been revealed to us—that insects also have various states of health and lack of health. In order for them to be relatively healthy, they need digested components. Their food sources need to be as such. But if their food source is not that good, or if it's missing something, they can suffer. Insects have died in the field. I've seen it many times. Sometimes people try to keep them as pets, and they die. We see them dying in the field all the time. They are susceptible to disease. They also have states of health and lack of health that are somewhat analogous to what you would find in a human.

If you're working with mosquitoes in the laboratory, and you try and infect them with the particular disease that they're supposed to be infected with—*Anopheles* mosquitoes with malaria, for example—you will not get a 100 percent infection rate. That is amazing. This is in part because when you're raising them in the laboratory, they're coddled. They're given everything they need. Therefore, they're in pretty good shape, and it becomes a little difficult to infect them. Out in the field, they might be more susceptible. Ones that are more susceptible might die. Others would be more susceptible to actually getting a disease, like the malaria protozoan or the Zika virus or anything of that sort. It can actually take hold of the system because the insect doesn't have the immune system to take care of it. It kind of sets up shop inside the insect.

JOHN: Absolutely intriguing. In all the various research and the different things that you've looked at through your very rich career, what has been something that really surprised you?

TOM: I guess the first thing that comes to mind is our olfactory research. We deciphered the insect olfactory code, and there are a massive number of details to be worked out. Over the past few months, I've talked a lot with my laboratory mates and others about how different insects are smelling the same molecule. This is something in the forefront of my mind right now.

I'm looking at frequencies right now for the Indian meal moth, trying to understand how they're smelling and how they're picking up pheromones, and seeing the differences between the various little receptors, gustatory receptors, ionotropic receptors, and things of that sort.

Putting this stuff together into a cohesive smell is something that has taken me off guard. I was under the impression twenty years ago that when you smelled an odorant, we had a receptor. That's the way I looked at it. We had a receptor, and this receptor responded to this particular smell. Now we know differently. We know, for example, that there are between two and three CO_2 receptors for insects. They don't have just one receptor for carbon dioxide.

When you take a look at this, you realize, "Wow, this is amazing." There's a complexity about it, but there's a simplicity too. It's more complex because it wasn't what I thought, but it's simplistic at the same time. It gives you a window into what God has done, and you understand that you don't know anything, where you think you did. Twenty years ago, when I started the laboratory and I graduated, I had a big, bad PhD. You think at that point that you know quite a bit. Now I can absolutely tell you how little I know. I could not tell you how little I knew twenty years ago, but I can now.

Because these things are not what I expected, I just have to follow the evidence. Being a scientist, you have to follow the evidence. You follow reason as far as your mind is allowed to. This is what has been given to us. We scientists are able to reason through things. Every human being is meant to reason. We are obligated do so in the scientific world, and then taking scientific evidence and applying reason to it in order to rationally wade through the muck—all the knowledge that we don't know.

Finding out these things has been part of the joy of being a scientist. It's part of the joy of being in a laboratory and coming up with new things. This is also part of the joy of speaking to groups at conferences and divulging some of this information—to bring the wonder of science, the wonder of insects, or even the wonder of insect olfaction to people who never had a chance to think about it before.

JOHN: Absolutely. We live in an incredibly beautiful world—if we take the chance to observe it. What you're describing is really fascinating for me because, in essence, you're describing how insects can be in the same environment but that they perceive their environment differently from an olfactory perspective. They're smelling things differently, which is really intriguing because we've known that to be true for various organisms from a visual and auditory perspective. From a visual perspective, different organisms can look at something and perceive different wavelengths, and so they see a different image. What you're describing, in essence—if I'm understanding it correctly—is that they are perceiving their environment differently from an olfactory perspective as well.

TOM: They are.

JOHN: That is fascinating.

What do you believe to be true about modern agriculture that many other people don't believe to be true? Or, rephrased, what do you believe to be true that a tomato grower in California doesn't believe to be true?

TOM: Insects are only attracted to unhealthy plants. This has become my most popular talk. You just get shock from people. You have spoken heresy. I know where you stand on this, John, and I understand many other people do agree with this philosophy. And I get that, but it still causes shock when I give this presentation. It's one of my opening slides.

If you believe that insects are attracted to unhealthy plants, your whole thinking changes on insects. Suddenly, you have no use for insecticides. It just follows with that level of thinking because you realize, "Well, I'm not competing with insects. They're just eating some of the garbage plants that I shouldn't be eating. So I really don't need to spray them anymore."

Now all the organophosphates and synthetic pyrethroids and carbamates and neonicotinoids—they're all unnecessary. As you reason through this, that's one of the conclusions you come to. You have to come to it when you believe that insects are only attracted to unhealthy plants. I don't even have to come out as an antipesticide guy. I can simply say that insects are not attracted to healthy plants. By extension, I would say, yes, insecticides are unnecessary.

Under specific circumstances, they have their role. I will be the first to admit that I've had fire ants. Sometimes they come into the house. I do have a Raid can in my house. We have had situations where the kids would leave food crumbs around, and the situation had to be taken care of. I'm not afraid to use insecticides, and I understand that they have their place.

Having said that, when I'm raising crops—and I do plant some stuff in my garden at my house—I'm never ever spraying it with anything. Not pesticides or insecticides. I don't have a knowledge of all the insecticides, but I don't use any of them. I'm not using any herbicides or fungicides. I'm just not going that route. I use insects as indicators. I go out and check my corn plants, for example, and I look to see if they're being attacked. If they are, I determine what insect is attacking them, and I figure out why they aren't healthy.

For example, the first time I started planting corn, I did have a few insects that were attacking it. By the third time that I was planting corn, no insects were attacking it, but I suspect deer came in and cleaned me out. This is the difference between the insect digestive system and the mammalian digestive system: we have a higher-level digestive system. We can handle healthy food. The deer are more interested in healthy crops. They're not going to go after unhealthy stuff. They'll leave that up to the insects.

But if you do have a healthy crop, you have to watch out for the deer. They cleaned out about 140 corn plants that I had planted for the benefit of my family, and I lost 99 percent of ears overnight. In a sense, I'd like to pat myself on the back and think, "Okay, well, at least the insects didn't get it." I'm an entomologist. That would have been embarrassing. But the deer sideswiped me and cleaned me out.

These are the things that you learn as you're planning and observing. You need to go out and take a look at your crop plants to understand what's going on. That will give you an idea of what you need to do in order to take care of the problem. There are many different ways of taking care of all of the problems that farmers run into when planting crops.

JOHN: What is a book or a resource that you recommend most often for growers?

TOM: I haven't recommended a book or a resource to a grower in a long time, to be honest with you. I read mainly academic books. I try to always look for the cutting edge type of thing. We scientists—we like to make things more complex than they actually are. This is something that I haven't been able to shake. Because of that, I probably am not well read enough to recommend a book, a publication, or anything like that in order to help a farmer.

The thing I really love to do is to talk to farmers directly. I like it when it's not filtered. I get a high from that type of thing because I can talk to them directly about their crop plants. Rather than just giving them a cookbook—what they need to do, which is usually what you find in a book—I prefer to actually talk to them directly because each one has their own story to tell and their own problem as well.

PASSIONATELY BREEDING *for* HEALTH *and* QUALITY

ED CURRY

When we were trying to increase yield, the limiting factor was the xylem and the phloem—nutrient transport. I call it the school bus. If you don't have a school bus, you can't get the kids to school. That transport system is everything.

Back before we understood anything about regenerative agriculture, by accident, we proved that if we could raise boron levels, we could raise calcium levels. If we could raise calcium levels, we could solve—or certainly minimize—blossom end rot.

The cell structure in a vegetable improves the shelf life. When we get the soil health right, the fruit density is way better.

Fertility is everything. Peppers are one of the most difficult crops to grow. You just name a disease and they get it. They get blossom end rot. Peppers are a struggle from day one. If you grow a good crop of chili in New Mexico, Arizona, Texas, California, you've done something. We've always tried to encourage our growers to improve fertility.

PASSION ISN'T everything in farming, but having some certainly helps. And Ed Curry has passion in abundance. A passion for improving pepper yields while improving disease resistance. A passion for breeding peppers with just the right amount of heat. And a passion for expanding the pepper industry in his native New Mexico.

Simply said, Ed Curry is an amazing grower. He may not have the academic credentials of some of the other interviewees in this book, but he's just as knowledgeable and influential as any of them. Through Curry Farms and Curry Seed, he produces 90 percent of the seeds for green chilis worldwide. He is the only remaining open-pollinated chili seed breeder in the world. He has been breeding chili peppers with a Mendelian approach, phenotyping plants by eye—not via a machine. And he's specifically breeding for a number of different quality characteristics and working to produce extremely high-quality seeds.

Developing high-quality seeds is a topic that is very important. It's something we haven't focused on enough—collectively as growers or as a culture.

JOHN: Ed, what are some of the key experiences that led you to where you are today?

ED: I'm a dirt farmer, and I love what I do. I was trained by the best—Mr. Phil Via, Ben Vialone, Everett Wood. I've learned under several really good pepper geneticists.

The one thing we all had in common was that pepper breeding was our life. My growers that I sell seed to, they take off elk hunting and deer hunting, and I always tell them, "You boys go hunt. I'm going to be looking for better chili." That's literally what I've spent my life doing.

My mother and father moved from Oklahoma in 1952. They had just about starved growing dryland peanuts and cotton, and they wanted to find a better way—a better place. They found Silver Springs Valley in Arizona. We're a closed-basin, irrigated farming system. We now use pivots and drip, but when my dad first moved there, it was all furrow irrigated.

In 1957, my dad planted his first crop of chili. I was born in 1956. So our family's been in chili virtually my whole life. In 1957 he didn't get a stand, and so he really didn't have a chili crop. He was determined in 1958, and he planted again and made his first chili crop. It wasn't great, but he made it.

Chili has one of the most diversified set of gene pools out there—a little over thirty thousand genes. We select straight by the eye. I look at plant habit. I look at fruit shape. I look at fruit size. We even look at little things—how the bloom sets, where the stamen is, where the pistil is.

Phil Via, who trained me—we lost Phil five years ago—when I was a very young man, we were standing in about sixty acres of chili, and he said, "All right, find me a hundred plants that are hot." We were in a mile-long field. He wouldn't let me taste them or lift the leaves to look at the chili. I had to do it strictly by eye.

I stood there and I got ninety-eight out of one-hundred. When we left the field that day, Phil said, "You get it." I'll never forget that. His point was to look at the small things. He had taught me about leaf shape, leaf size, pointed leaves, kill-type leaves, and many different things.

Doesn't that fit in life as well? My dad always said to take care of the little things, and the big things would take care of themselves. Well, in plant breeding, it's certainly true. I've developed a team through the years that works with me there on the farm. Everybody on my team has been with us many, many years—over twenty. The idea here is that we love what we do. We grow our entertainment. We don't have to go hunting! We don't have to leave the farm. We just have to leave the house and go into the first field on the farm.

Something you've taught me is how important nutrients are to seed development. I'm always searching for better embryo development. In my chase of better seeds and better genetics, I spent several years farming in Mexico. I learned a lot. We went down to the coast—about five hundred miles south of home, right on the ocean. The reason I went to Mexico is that I could gain a generation. I could harvest in October in Arizona, and then I could take that seed and plant it right away in Mexico and get another generation and then come back home. That was the whole idea.

There were several differences in Mexico. We were growing in wintertime, so there were short daylight hours, and this changed how the fruit looked. It was really hard to phenotype. As I look back, I realize it was God's way of bringing my educational process into one. I started realizing that phenotyping was much different than at home.

My biggest customers are in and around Hatch, New Mexico—Deming, Las Cruces. We have a pretty good presence in California and in Texas and Colorado. Not to mention that about 40 percent of our sales are in Chihuahua, Mexico, plus a little bit in South Africa, some in Chile, and a little bit in Peru. One thing Phil taught me was that we have to have a little bit of variability. If we hone it down to where it just grows good in our valley, it's not going to do well in other areas. So we always look for a little bit of diversification in plant habit.

But if we've got a little bit of diversity—that's the fun part of open pollinating.

JOHN: Your farming operation has really shifted and changed, even in the short time that I've known you. Can you tell us a little bit about your actual farming operation, your cultural management practices, how you seek to produce—the types of things that you seek to do to produce—really high-quality seed—and how that has changed in the last decade?

ED: That's one reason I mentioned our time in Mexico. I knew early on that good fertility meant healthier seed. I'm selling to growers that are direct-seeding. Very few people transplant chili. Some do, but as a whole, we're direct-seeding, because in the Southwest, it's so brutal. We get hard spring winds; we get all kinds of difficult weather, including freezes. Most people don't know that chili will take about a twenty-seven- or twenty-eight-degree morning as long as that seedling is really healthy. If it's not healthy, it won't take it. If it's healthy, it'll take it.

I knew early on that we needed healthy seed. I'm always looking for a better way to increase vigor and germination so that my growers start with an advantage.

I will tell you this about chili: in the Southwest desert, you know what you're going to make within the first month. I know that sounds a bit crazy, but sometimes, I've seen it take up to thirty or thirty-five days to get a seedling out of the ground. Anytime it takes that long, I know that grower is not going to have a really good crop. He'll make chili, but he's not going to make the best chili.

On the other hand, if we get it out of the ground in ten or fifteen days, we're going to make a crop—at least we've got a good chance.

JOHN: I think that's a very interesting point, considering the fact that I've had conversations with grain crop producers in the Midwest the last number of years who are attempting to germinate corn seed in an optimal greenhouse environment, and it's taking fourteen to twenty days to germinate. I think that's a problem for a corn seed. We have had such a substantial deterioration of seed quality within the industry at large. I'd love to hear a bit more about some of the specific things you're doing to try to increase seed quality—to speed up that germination time.

ED: Some of the in-furrow products you brought us do really well. Gary Zimmer talks about having healthy biology on your seed. We have definitely found that biology on the seed, early on, gives us a much better start. The first thing we saw in AEA [Advancing Eco Agriculture] products, in the beginning, was ten days earlier fruiting. Ten days is huge. Anytime you've got earliness, you've got better plant health—every time.

JOHN: Ed, what are some of the moments in your journey that have led you to seek to make a change in your farming operation?

ED: Well, that's my mental condition, John! Seriously, though, five and a half years ago, I had a precancerous situation in my esophagus. Sometimes I've wondered if it was from tasting so many hot chilis because part of our work is in heat, and we genetically altered capsaicin on a couple of our varieties. We call it the "mean heat." It attacks.

I feel almost certain that's part of what caused my trouble. And by the way, I'm not the only breeder I've talked to who's had health problems from tasting fruit and doing certain things. It's kind of interesting. I've had several come to me and say, "Yeah, I got in trouble doing that."

JOHN: Specifically from working with genetically modified varieties?

ED: Well, in this case, it's just open pollinated. But it was the capsaicin, I think. I can't prove that, but I sure think it was. David, who works with me, started having trouble as well. We're tasting these things until we can't taste anything. Capsaicin will deaden the nerve endings. I guarantee you that my tongue has experienced more hot chili than I would ever want to explain.

By the way, capsaicin is becoming a very big industry. It can be used for local anesthesia—you all already know about the mace products and some of those. The capsaicin field is just beginning. Twenty years from now, it's going to be used for a lot of good things.

So I ate my esophagus up. They did what they call a "fundoplication" on me, and I had a precancerous situation. They took part of my stomach. They made me a new sphincter valve, and my health declined steeply. For the next three years—at one point I got to where I could hardly get out of the house. I was so weak and so allergic to everything.

I found a holistic doctor in Scottsdale—Dr. Rob Ellsworth. Rob started giving me food allergy tests. We did heavy metal tests. We went down the line. I'm telling you the truth: if I would spray my front yard with Roundup—just a little bit—I would get sick.

JOHN: Lucky for you!

ED: We started figuring it out. I have a personal investment in understanding health. I had to figure out my own body. I was deathly ill. I got over the surgery—I got over all that—but a specialist, a friend of mine, a doctor that I work with in the chili business, told me that my ability to digest B vitamins was extremely impaired.

I started looking at different forms of vitamin B to digest. I started looking at my own farm. What was I doing? We had already started to go organic, and today on my farm, we have five hundred acres established as certified organic. The reason was my own health. I poo-pooed all the stuff about Roundup being bad until it made me sick. And then I said, "Whoa, hold it. There is something to this." And there is.

Ironically, you know what it is for me? It's the carrier. It's the carrier in the Roundup.

JOHN: It's actually more toxic than the glyphosate itself.

ED: Exactly.

JOHN: You mentioned just now that you've been in the transition to organic for a period of time. One of your major reservations about becoming an organically certified farm was the challenge of disease control. Can you describe diseases in chilis and what's been happening?

ED: THIS is the challenging part: disease. I was scared to death of it. Most folks understand what *Phytophthora capsici* is—if you've grown any squash or any soft vegetables. I've talked to doctors all over the world about this, and they're shocked. "You're in the desert. How could you have *Phytophthora*? How could you have bacterial spot—*Xanthomonas vesicatoria*? How could you have these things? They're all water related."

When we get a rain, we get a real rain. In both 2016 and 2017—a week apart in July of both years—we got three and a quarter inches, with hail, in about forty-five minutes.

We farm on a slope. The water moves. So do *Phytophthora* spores. And *Phytophthora* spores sporulate. I describe it like a wheat fire. The wind will direct where that fire goes. That's how *Phytophthora* spores move. You get the right temperature, the right water, the right moisture, and the right wind, and it will go. You can follow it across the chili field, just like a fire in a wheat field. It destroys the fruit, and then eventually it gets to the roots and destroys the plant.

Our first experience of that was 1985. I had a beautiful crop that was doing really well, and we were harvesting. I had water, and in those days, we were fer-irrigating. We always water ahead of harvest so that the fruit is good and fresh and hard. We were watering ahead, and we got to the twenty-third of August 1985. We get storms off the coast. We're not really that far from the ocean as the crow flies—about three hundred miles.

It started raining and drizzling. Just slow—no sunshine—which is rare for the desert. But it happens every year at one time or another. And that *Phytophthora* went crazy. I went from harvesting seventeen tons to the acre—which was a super yield in those days—to skipping the next fifteen acres. I didn't even harvest it. It was all gone.

What little was there was so hard for the cannery to sort through that we just had to give it up. I was a young man with several children. I thought life was over. How in the world were we going to make it through that? We had just lost fifteen acres. In our business, fifteen acres is a lot because the inputs are high. Today we get about six thousand dollars an acre. In those days, it wasn't quite that high. But it's intense agriculture.

I was just wilted. We've battled *Phytophthora* for years: 2010 and 2011, consecutively, we lost half of three hundred acres, both years. It represented a little over a million dollars of input. And when I say "lost," I don't mean, "and we went in and harvested some." Out of 300 acres, 150 acres were gone. There was nothing to harvest—even in our seed blocks. We were able to find a little bit to continue our genetic work, but only with lots and lots and lots of searching.

I'd been on the biology track a little bit—better health, better soil health, etc. When we found regenerative agriculture, we started having hope. In 2017, the only field we completely lost to *Phytophthora* was my conventional field. My organic field did have some, especially the draws where the water runs. But overall, it was pretty good. My conventional fields—in 2017—out of thirty acres, we probably harvested the equivalent of about four or five. That was using Ridomil—all the high-powered stuff. This past year, I didn't have any conventional. I have found that soil health gives us a better chance.

Also with bacterial spot, that's where I really had the most arguments from scientists. The university people say, "Well, you shouldn't have bacterial spot. You're in a dry climate." Do you know what happens to us in the spring? It's not uncommon about the tenth of May—then along about the eighteenth or the twentieth, about every ten days—we get those cold fronts that come down. We get really voracious winds. One day, we had 110-mile-an-hour wind. You get sand moving. It opens places for infection.

Bacterial spot—*vesicatoria*—is all over our desert. There are all kinds of host plants native to Arizona. When that wind blows, those spores are in the sand, and it inoculates us. Does it get sick right then? No. Give it a little bit of time. Give us some spring or some summer rains, and here it comes.

The thing about *Phytophthora* is that I know where I have it. But bacterial spot, you don't know till it's really ravaged you, and it just keeps coming and coming and coming. Finally, you can't set any fruit, and you've got this really sick plant that you can't do anything with. And since we've been organic, have we whipped it completely? We still battle it, but is it better? Yeah, it's quite a bit better.

JOHN: You mentioned a moment ago how yields have changed today from where they were thirty years ago. I think this is a trend that we see with a lot of crops—as crops become more developed and farmers manage them more intensely, we begin tapping into a much greater part of the genetic potential that many of these seeds already have. How have yields shifted and changed in the time that you've been doing breeding work?

ED: That's the fun part of our story. We want to spread the green chili industry—which is basically Hispanic Mexican food—to the North and to the East.

But you all are meat and potatoes people; you don't like spice. So what was one key thing? It couldn't be hot. The old 64 variety that was bred by Roy Nakayama in New Mexico State University—it's great variety—but the old tradition was to keep your seed year after year.

If you had one hot chili, pretty soon, you had several hot chilis, and pretty soon, you had a lot of hot chilis. It was very common that you would be eating a can of chili, and two or three of them would be mild, and then—boom—one of them was really hot. And that's what we couldn't have.

What put Phil and me on the map was that we were the first ones to stabilize heat. We were able to go to the cannery with a better yield and better heat control.

When we started, the average yield in green chili was seven to eight tons across the board. The first year we released one of our varieties, which then was called Arizona 20, it jumped that average yield to fourteen tons—the first year. Well, that got everybody's attention—almost double.

It's not uncommon now for us to touch thirty-five to forty tons. Average for the industry right now is eighteen to twenty. I talked to one of the guys at the cannery yesterday, and we're going to start another variety that's reached thirty-two to thirty-six tons with one pick. Oftentimes, they multipick and reach much higher levels.

When we were trying to increase yield, the limiting factor was the xylem and the phloem.

JOHN: Nutrient transport—the nutrient-transport pipeline.

ED: I call it the school bus. If you don't have a school bus, you can't get the kids to school. That transport system is everything.

I won't publicly say what I used because all my competitors will try to put it in their hybrids, but we went to one plant that I knew, in an annum crop—in a one-year crop—that would grow a stem about an inch and a half, sometimes as large as an inch and three quarters around. I knew that it had a better vascular system. I knew it had a bigger pipeline.

We crossed that with one of our chilis. We crossed it and then started selecting. And to this day, we're still working with many of those crosses. That's where we got the better yield. Also, way back before we understood anything about regenerative agriculture, by accident, the guy who was working with it was doing our sampling. We proved that if we could raise boron levels, we could raise calcium levels. If we could raise calcium levels, we could solve—or certainly minimize—blossom end rot. All with just the application of boron.

All the Western handbooks and all the studies said that too much boron would be deadly, right? Everybody knows that. One time, I had a problem with a pivot: we were running boron, and the pump didn't shut off. We pumped a whole tank of boron onto one spot. We watched it, and we took samples the rest of the year from that spot. We didn't kill any chili. We stunted it a little bit. But guess what? The calcium levels were better all the way across, even with extreme boron rates.

The same thing happened to a farmer in Sonora. He got way too much boron in a drip system on jalapeño, and that block of jalapeño was better than any of the rest. So we all started increasing our boron applications.

Then we started comparing with the lettuce growers. I used to grow and ship a lot of lettuce. I started comparing what we were doing with the lettuce boys in Salinas, California. We noticed the same thing—this was in the eighties—lettuce with higher boron levels had higher calcium levels. And guess what? We got better, healthier lettuce heads.

JOHN: Some of that is getting to be very well understood. I find your story about breeding for vascular tissue and breeding for stem size to be really revealing and intriguing. Bruce Tainio, who is no longer with us, was a seed breeder and a geneticist before founding Tainio Technology. He developed some of the foundational wheat and soybean germplasm that is still being used today as foundational seed stock because he focused on developing varieties with really large and aggressive root systems, which still—to this day—have not been exceeded.

Within all of this, Ed, what is something that you believe to be true about agriculture that many people might perhaps not believe to be true?

ED: That's a good question. I spoke at a funeral this past Saturday for my neighbor. I rented his farmland. His son and I grew up together—a super family. When I was talking to the family about stories to share, his son said, "You know, when I was very young, in high school, Dad held up a handful of dirt.

And he said, 'Son, in this dirt lies all life.'"

From where I sit, and from what I've experienced with my own health—my own esophagus damage—I've realized that all the chemical farming—all the things we thought we were doing right—when we started using commercial fertilizers, it wasn't healthy for the soil. If we tear the soil down, we're going to have less fruit quality.

I can give a whole talk about cell density in fruit. Out of our single plant selections, we found a row that didn't go down when it froze. In everything else, the tissue was broken down. This row it wasn't. That was genetics in that particular case. We've kept that line, and we've been working on it; in fact, we're really proud of that line. It's all about cell density and cell structure in the fruit.

If you go to a store and you buy an expensive pima cotton sheet, it's one thousand threads per square inch, right? That's a quality sheet. If you go to Walmart and buy the cheapest thing you can, made in China, it will have three hundred threads per square inch.

Vegetables are no different. The cell structure in a vegetable improves the shelf life. When we get the soil health right, the fruit density is way better. We look at so many chilis. I cut chilis day after day after day. We look at them, and we study them, and we carry little magnifying glasses, and we do all kinds of things to see the little things. That's the fun part. It's one thing to look at a set of genes and to look at alleles and DNA on a computer screen. But if you study the little things, they will talk to you. I needed to carry the same things out in raising my teenage son—I needed to study the little things sometimes a little better!

When we get better health in the soil—in the plant, in us—we are going to help our customers be healthier. Did you know that one chili has over double the amount of vitamin C of one orange? That's not counting vitamins A, D, or E. Chili is an amazing plant.

Pepper is so diverse in so many ways. I'm glad God made it that way. You know why? Because there's no gene shooting going on in chili. They've tried time after time—150,000 or 200,000 shots—and maybe one will take, and then it won't take off. But that's okay.

JOHN: That's perfect. It's better than okay.

Ed, what are you working on right now that you're really excited about? What do you still desire to experience on your farm?

ED: Probably the thing I'm most excited about are our newer, higher yields. Our growers struggle to compete with Mexico—right across the border. Everybody thinks of New Mexico as being the kingpin of chili, and it is. We've done a great job of marketing that, and I'm proud to be a part of it.

But today in New Mexico, there are 7,000 acres of chili for canning purposes. In Mexico—in Chihuahua, the state bordering Luna County in New Mexico—there are 92,000 acres. If you go a little bit further down into Zacatecas, which is the largest pepper-growing state in the world, there are 250,000 acres—a quarter million.

So when we look at what we've held on to in the States, it's very little. If I'm going to hold on to my growers, and keep this industry in the state, we have to have higher yields. We have to give growers an advantage.

So yields are number one. Disease control is number two. And also flavor. We've got a chili that we call Grandote, and it has a dark-red cherry flavor. And then we've got the yellows. One test showed that one yellow variety had one hundred times the nutritional value of spinach. That means that you would have to eat one hundred ounces of spinach to equal one ounce of this yellow chili.

We're going to change the world. We're going to help people be healthier. That's exciting to me.

JOHN: That's awesome.

So, Ed, you've had four decades of experience as a grower. What advice do you have for other growers? Actually, not so much for growers. I think there is a strong desire by many in the vegetable production space to develop stronger, more stable genetics. What advice do you have for a young person who wants to learn more about breeding and developing their own varieties?

ED: Come to the farm.

JOHN: You might be inundated.

ED: I love teaching young people. Phil Via passed it to me. I have a high school degree. I didn't go to a public university. Now, I lecture at universities! But I don't have a sheepskin, if you will.

JOHN: What resources would you recommend?

ED: You know, I like Luther Burbank. He's inspired me so much. I'm going to go study Luther a little bit more—some of his early information on plant health.

We took a wrong turn, and we forgot about soil health. What did we forget about in breeding? I'm not saying all science is bad; some of my good friends work at New Mexico State and UC Davis and the University of Arizona. But I like what you said about going back to healthier root stock. History will teach us. What did people in the past learn? What did they know? Somebody knew something about soil health hundreds and thousands of years ago.

JOHN: I've done consulting work around the world, and there aren't very many people left who have the base of knowledge and awareness that you do on open-pollinated seed breeding. It's not common knowledge, and I think it's something that's very important and very valuable for us to tap into.

Here are some questions other people from the (live) audience have for Ed.

Audience Question: I was interested in your story about the trial of picking out the hundred chilis. What immediately came to my mind was Luther Burbank. Are you seeing a difference in the expression of genetics when you have improved biology in your fields? In school, we're taught that abiotic stress creates more of an expression in genetics. But in my experience the opposite is true—the more you see balance in the soil, the better the genetic change. Are you seeing that? Are you seeing a difference in what you see in the genetics now—now that your soil microbiology is changing?

ED: That's a good question. I strongly believe that genetic expression—what turns genes on and off and what makes certain things happen—I think we've got a lot to learn there. But the one thing I see for sure is that the healthier we are, the better genetic expression we get.

People ask me all the time if you can change how hot a chili is by fertility, by water, or whatever. The answer is absolutely. Capsaicin is the protective mechanism in the pepper. That's all it is. When a pepper stresses, the capsaicin goes up because it thinks it needs to preserve itself. So the pepper is hotter. So, to your point, the healthier the soil health we have—and if we're in balance—the better.

My good friend, grower Gary Shiflett—he's not organic, but his farm is balanced. He was recognized as the national vegetable grower of the year about ten years ago. He's an amazing farmer. You walk on Gary's farm, and it's balanced.

I firmly believe that we see different genetic expressions through that. Heat's the fun thing to talk about. With pepper, if you get to the harvest and you've got hotter chili than normal, then you either missed water or you missed something because it's going to tell you.

QUESTION: Have you been keeping track of people that use your method of breeding in other crops—corn or zucchini or other plants? Are there growers in other regions that use the same method—who, instead of the lab, just grade by eye?

ED: I really don't know a lot of breeders anywhere who do it. There was a fellow in Hungary that did that for the yellow banana peppers and the little yellow and red chili—cherry peppers. He did almost exactly what I'm doing. Sadly, he had a bad stroke and became completely impaired, and that whole seed lot was lost. In other vegetables, I'm sorry to say, I know hardly anybody.

If big companies come to me to look for parent material—even if they're going to do hybrids, they've got to have parent material. They've got to have the set of genes to work with. You can't build a Lego set without the Legos. In answer to your question, universities, by and large, are not teaching young people what I do. Even if we're going to continue with hybrids, and all the other work that's got to be done, somebody still has to build the gene pool.

That's what we're doing. That's the fun part. We're building gene pools. I'd love to encourage young people to get into more of this.

Question: There are problems with citrus greening in bananas and with coffee. But it seems to me like we're trying to adapt to poorer and poorer soil and environmental situations. The first thing we do is say, "We need a new breed," or "We need a new spray," or something. But we're not dealing directly with health.

What happens with conventional breeders? Do they tend to breed so that they don't need the trace minerals that might not be in the soil anymore? Is that what happens, or are they actually trying to breed a stronger plant? What kind of thinking goes into developing a new breed? Breeders are trying to do this with other plants right now and are failing, and I think part of that is just a nutritional problem.

ED: We have never tried to breed something that would do better with less fertility. I have tried to breed for drought tolerance, though. Without water, stress is induced.

Fertility is everything. Peppers are one of the most difficult crops to grow. You just name a disease and they get it. They get blossom end rot. Peppers are a struggle from day one. If you grow a good crop of chili in New Mexico, Arizona, Texas, California, you've done something. We've always tried to encourage our growers to improve fertility.

How horrible would it be if we bred something that didn't carry the minerals into the fruit so that we get those nutrients? That point needs to be hammered home. I so strongly feel that our health across this world has to do with good genetics and good fertility.

WATER *and* NUTRIENT MOVEMENT *in* PLANTS *and* CELLS

DR. GERALD POLLACK

Life is water dancing to the tune of solids.

—ALBERT SZENT-GYÖRGYI

This amazing book has changed my understanding of all the processes going on in water which I was confident I knew about—the understanding that dictated my many years of teaching and organized my research. I must now come to terms with the demonstration that water is not just a medium in which physics and chemistry happen, but a machine that powers and manages physics and chemistry.

—MARTIN CANNY, regarding *The Fourth Phase of Water*

We found that if you add infrared energy, that this flow through the capillaries—which keeps going even when the heart is not beating—amplifies enormously. We're confident that the phenomenon of water flowing in a tube applies not only in our cardiovascular system but also in the vessels of plants and trees.

The concept that you must have an impermeable membrane around the cell is not correct. I think if one looks objectively at the evidence instead of merely accepting what textbooks like to say, one would come to that same conclusion. Substances can get absorbed by the cell because there is no impermeable barrier.

DR. GERALD POLLACK is a leading researcher in the area of water structure and water ordering—or as he refers to it, "EZ" (exclusion zone) water. Jerry's research at the University of Washington has won some exceptional accolades and is considered by some to be part of the most important and groundbreaking research of the twenty-first century.

Jerry's research into how water functions in plants and in cells has tremendous implications for how we understand plant nutrition, nutrient absorption, nutrient flow within plants, and—very importantly—how we understand the absorption of nutrients and compounds by plant root systems and by plant leaves.

Any citizen-scientist, farmer, or agronomist who wants to understand how to manage plants and livestock and human health should get two of Jerry's books: *Cells, Gels, and the Engines of Life* and *The Fourth Phase of Water*. They strike at the very foundations of how we understand biology, and their ideas will fundamentally shift and reshape our understanding of biology in the near future.

JOHN: Can you provide an overview of exclusion zone water and how it might function in plants and living organisms?

JERRY: The quick synopsis is that we've all grown up learning that water has three phases. There's solid (ice), there's liquid, and there's vapor. We all learned that. On the other hand, a suggestion was made by a famous physical chemist more than a hundred years ago that it's really hard to interpret the known behaviors of water with these three phases. You try to do it with the understanding there are only three phases, and you simply can't. It creates an anomalous behavior—"anomalous" meaning it doesn't fit the theory, as of today. There's a famous website that details sixty or seventy of them.[7] A chemist looking at these anomalies thinks something has to be wrong with our understanding of water. There can't be so many features of water that don't fit the theory.

---

7 Martin Chaplin, "Anomalous Properties of Water," Water Structure and Science, http://www1.lsbu.ac.uk/water/water_anomalies.html.

More than a hundred years ago, a suggestion was made that there must be another phase of water beyond those three—a phase somewhere in between the liquid and the solid. This idea has been kicked around for some time, particularly by a guy named Gilbert Ling. He's ninety-nine years old right now, and he's written six or seven books on the subject. Most of what he's written points to evidence that there is another phase. He doesn't call it a phase of water; he calls it "structured water"—that is, where the molecules are not bouncing around randomly and pointed in every which way, but they're ordered. They're like dipoles with a little plus and minus at either end. All these dipoles are lined up with one another, and he's argued, for many years, that water inside the cell is exactly like that. It's ordered.

Another person who had this idea was Albert Szent-Györgyi. He passed about twenty or twenty-five years ago, and he is considered to be the father of modern biochemistry. Having been in various fields, including water, he knew that the organization or the order of water is really important in biology. In fact, a famous quote of his is, "Life is water dancing to the tune of solids." He knew that water was centrally involved in everything to do with living systems.

The fourth phase of water is exactly the kind of thing that these people were thinking about. We've found clear experimental evidence for it. It's a phase of water that is indeed in between a solid and a liquid. To appreciate the consistency, think of raw egg whites. It's gel-like—like a loose gel. In fact, it's the main component of water inside of cells. It's an ordered phase in which the molecules are lined up a bit different from what Gilbert Ling has put forth about dipoles organizing themselves. It has a different structure. And unlike ordinary water, which is neutral, structured water usually has a negative charge.

It's built by energy. You need some kind of energy to build order, and this is ordered. The energy actually comes from light, particularly infrared light. Infrared light is all around us. If you were to turn off the lights in a room and pull down the shades so that it's completely dark, and if you were to then take out an infrared camera and create an image, you'd get a beautiful image because everything is emitting infrared energy. It's used as a nightlight, especially for military purposes.

This energy is all around us. It's free for the taking. This free energy, which is all around us, builds this kind of water. We found that it forms when water meets a hydrophilic surface. Hydrophilic—water-loving—means that if the surface of a desk or something has a droplet of water on it, the water spreads out. This is as opposed to hydrophobic—water-fearing—in which case the water beads up and shirks away from the surface because it doesn't like it. Usually, for most hydrophilic surfaces—not all but most—when water meets these surfaces, the water transitions from ordinary H_2O to this fourth phase of water. The reason we call it the fourth phase is because it's just different from ordinary liquid water and certainly different from solid water, although the structure is somewhat similar to ice. It's not the same, but there's some resemblance.

When water meets such a surface, what happens is that the water layers—starting with the first layer right next to the surface—undergo transition from ordinary water to fourth-phase water. It's not just a single molecular layer. That would not be so interesting. It turns out that many, many, many molecular layers build in that way. We found that up to millions of molecular layers can build from such surfaces, all powered by the energy from light—and especially infrared light, which is more powerful to build this kind of structure than other wavelengths of light.

Because it's an ordered phase—because the molecules inside of this phase are not bouncing around randomly—they're well organized and ordered, in a sense. It's like a crystal. It's a liquid crystal. If you can imagine a crystal of, say, ice—which is solid crystal—what happens is that as the ice builds, it excludes all kinds of materials and particles from the ice. Otherwise, you wouldn't get a pure crystal of ice; you'd get ice containing some junk or garbage or whatever. As the crystals build, in order to gain the purity that they show, they have to push out everything else. They exclude the other materials. This is exactly the same way the fourth phase builds: it pushes out substances. That's why we call it "exclusion zone"—because it excludes.

There's another reason for calling it an exclusion zone: because "EZ"—exclusion zone—is easy to remember. The only problem is that it works in the U.S., but when you go to Europe, it's "E-Zed." That's not quite as cool, but it works anyway. It has a couple of different names—it's "EZ" water (which is what I tend to call it); it's "fourth-phase" water. Some people still refer to it as "structured" water, although that really is not an appropriate term because everything has structure. "Ordered" water is another. They're all essentially the same term.

JOHN: There are a lot of implications for biological systems—to our understanding of how cells work and how nutrients are transported. There are two pieces in plant nutrition and agronomy that there isn't a good explanation for, that I know many farmers and many agronomists struggle to understand. The first is, How are nutrients actually transported? And also water—how is water transported inside a plant? Secondly—related to that—how are nutrients absorbed by root systems and then transported to the upper plant? We know there is plenty of evidence to suggest that the current models for how this is thought to work doesn't fit the evidence. We know the model is not quite correct. How might our understanding of EZ water be able to shift our thinking?

JERRY: This question always arises. How does the water get transported? The question is most interesting when you think of hundred-meter-tall redwood trees. How does a column of water get transported from the roots all the way up to the leaves on top? This is a conundrum.

In fact, sitting on my desk right now is a picture of Martin Canny. Martin wrote a famous *American Scientist* article called "Transporting Water in Plants"[8] in which he indicated how difficult the problem was to solve. A column like that is so high that the amount of energy it would take to raise something like that is immense. If you've got a head of water on top of you that's three hundred feet tall, that's a lot of weight. It's hard to imagine getting water to move up against such a heavy load. At the time, he was not clear on what he thought, but eventually—after many discussions—I think he was aligned with our thinking. He wrote a nice little capsule at the back of my *Fourth Phase* book. He said that by reading my book he had learned not just how physics and chemistry apply to water but that, through water, he had

---

8 Martin Canny, "Transporting Water in Plants," *American Scientist* 86, no. 2 (March–April 1998): 152–159, https://www.jstor.org/stable/27856981.

learned more about how physics and chemistry work.[9] I was really pleased with that comment.

So how is water transported? We think we understand the answer to this question because we discovered something that was completely unexpected. We ran an experiment in which we took a thin tube—one maybe five hundred micrometers in diameter, made of hydrophilic material, resembling a xylem tube—and put it inside a bath of water. The water contained a lot of particles, so that if there was any water movement, you could easily track it by looking at a microscope and watching how those particles behaved.

A young student working in our laboratory came running in. He said, "I put this tube in the water and I noticed that the water was flowing through the tube, just like water flows through a straw, and it doesn't stop. It just keeps going." He thought it was pretty important. (He was supposed to be doing something else. But most of the students don't listen to me. They find more interesting stuff than what they were assigned to do. And I actually encourage that sort of thing because some important discoveries have come from it!)

I agreed that this was extremely important because, usually, if you've got a tube and you want to drive water through the tube, you need to develop a pressure difference between one end and the other end, and that pressure gradient drives the flow. It's like a blood vessel, and you want to drive blood through the vessel. You need the heart to do the pumping. But in this experiment, there was no pressure gradient because the tube was sitting, horizontally oriented, in the chamber of water, so there was no pressure difference between one end and the other. Yet the fluid kept flowing through. But you need energy to push fluid through; you can't get something for nothing.

As he came in and delivered this news, I was thinking that if he were right about this, that it would be important because it would demonstrate what we had known up to that point: that the energy comes from the light that's absorbed in the water, especially the infrared light. We checked the experiment for artifacts or errors or such. It was correct. It demonstrated the importance of this kind of energy that's being absorbed from the environment.

---

9 Here's the full quote: "This amazing book has changed my understanding of all the processes going on in water which I was confident I knew about—the understanding that dictated my many years of teaching and organized my research. I must now come to terms with the demonstration that water is not just a medium in which physics and chemistry happen, but a machine that powers and manages physics and chemistry."

When you think about plants, the environment out there is pretty important. We did many experiments afterwards that showed that it's the energy from outside—particularly the infrared energy—that gets converted into a different kind of energy. Actually, it's a charge gradient, but we don't need to go into that at the moment. It's in the *Fourth Phase* book. But this is what drives flow through the tube.

Today we have a completely different understanding about the driving of flow through a tube. We can apply that, for example, to what happens in plants and trees. We have these thin tubes, and in this case, they run vertically. We know that the energy from outside can be used to basically push the water through this kind of environment. Martin Canny, by the way, did an experiment that I describe in my book. He took some xylem, and he put some particles inside, and what he saw was an exclusion zone at the edges of the xylem. In other words, this fourth-phase water exists as an annulus just inside the xylem tube, and this is exactly what we found in our experiments. We used the hydrophilic tube made of various materials, and always—right next to the material's surface or the tube, just the inside—is our EZ water, our exclusion zone—the same thing Martin showed exists in xylem with plants.

This reinforces the idea that what my student saw in the laboratory may be operative inside trees: the energy is coming from outside. Without going into detail about the mechanism, it has to do with charges that develop inside the tube and charge gradient. If, for example, you have a lot of protons, the protons first attach to water molecules to form hydronium ions. Those hydronium ions repel one another. If there are fewer hydronium ions up high and more down below, the hydronium ions—which are basically a protonated water—will move upward. We're pretty confident that this mechanism is a mechanism that applies in the xylem of plants. With a couple of diagrams, it's not so complicated.

A side point is that we think that this also occurs in the human cardiovascular system. I began my career studying—or simulating, you might say—the cardiovascular system's pressure and flow. I built a model and I was really excited about this model. I thought we had pretty much solved every problem there was to solve because we could explain the pressures and flows at different points in the cardiovascular system, and everything occurred as a consequence of the pumping of the heart. For many years, I was proud of that, although it made no impact. I began to understand why. Usually when people build models, their models are filled with assumptions. If you're not sure about whether these assumptions are accurate or not, then it plants a kind of doubt.

I thought I really understood the cardiovascular system until I went on one of my trips to Russia, to Moscow University, where my good friend Vladimir Voeikov introduced me to his neighbor, the guy in the next laboratory.

This guy was studying the cardiovascular system, and he tells me, through translation, "In the cardiovascular system, there's a very big problem."

"Oh, really? Tell me about it."

He said, "Well, the big problem is that capillaries are so small."

The system starts with the arteries and arterioles down to very thin capillaries, then the venules and the veins and back to the heart, and it circulates. But when you get to those tiny capillaries, there's a problem, because in healthy young adults, the diameter of the capillary is three or four micrometers—compared with the red blood cell that needs to squeeze through it, that's six or seven micrometers. It looks like a plumbing problem. It looks like you have to force those red blood cells somehow to get through these narrow conduits. It looks like a screw-up of Mother Nature. How is it possible? This is not a plumbing system that you would think Mother Nature would devise, but that's the way it is. Of course, there are reasons for it, obviously.

In order to squeeze those red blood cells through, they have to bend. They have to be squeezed and pushed through. So this Russian guy calculates the amount of resistance that would be encountered in squeezing these red blood cells and pushing them through. He comes to the conclusion that the resistance is so high that if the heart were fully responsible for driving blood through those capillaries, the heart would need to develop an extraordinarily high pressure—something like a million times the pressure that it actually develops. You might call that high blood pressure! But we're talking about six orders of magnitude. This is a problem. He says it can't be—there must be something else besides the heart that's doing this because it just can't be.

That had a big influence on me. The Russian researcher had a few ideas about mechanisms. I had my own ideas, and I knew about our work—where if you put a tube in a chamber of water, flow goes through the tube like through a straw. I'm thinking, "Wow! The same phenomenon could be at play in blood vessels." We tested it, and we found that was indeed the case. We found that, first of all, like others have found, if you stop the heart, the flow doesn't cease. This is the first thing that is shocking. Many people don't know about this, but there are a handful of studies over the years that have shown that if you stop the heart in any of a number of ways, you don't stop the flow. It goes down quite a lot, but it keeps going. How is it possible that it keeps going? There must be something that is at work in addition to the heart.

I think what's at work is the mechanism that I've been talking about, where you can get flow propelled by electromagnetic energy—infrared light energy. There's lots of infrared inside our body because of the metabolic energy of heat. And heat is where you get infrared energy. It's like going into the sauna. You get lots of infrared energy, and that energy is absorbed by the body. We found that if you add infrared energy, that this flow through the capillaries—which keeps going even when the heart is not beating—amplifies enormously. We're confident that the phenomenon of water flowing in a tube applies not only in our cardiovascular system but also in the vessels of plants and trees.

JOHN: Considering your blood example, we know that having the right electrolyte balance within blood is very important—sodium and potassium, for example. You mentioned the importance of having the appropriate proton pressure and proton gradient within plants. I'm wondering, how does this awareness transfer to our understanding of nutrient mobility within the vascular tissue of a plant? Are there important nutrient balances that we should be paying attention to in plant structure as well?

JERRY: There are, but I don't know them, and so I would hesitate to speculate. But I will say one thing: if you want to transfer something through tissue or through a vessel, there is a way to do this. I hesitate to come across as though I'm trying to sell my book, but in my first book, *Cells, Gels, and the Engines of Life*, this is described in detail. Imagine a tube filled with EZ water—with structured water—and that tube tends to exclude almost everything. We can't say everything because we haven't studied everything. However, if there's a region along the tube that is not "ordered" or EZ water but is ordinary water, the solutes or nutrients or whatever you like will tend to gather in that zone.

There is a process that I describe called "zone melting" that pretty much does exactly this.

Along the tube—inside the tube—you've got structured water, but in some region along that structured water, let's say you have a zone that's not structured. All the nutrients would be there. Now, if this nonstructured zone propagates along the inside of the tube, then the nutrients will move with it. This is a way of transporting substances, and I think that this may operate inside the cell. If you think of the cell as containing only structured water—and I think there's clear evidence that most of the water inside the cell is EZ water—then all the nutrients would seem to get excluded, and that won't work. You need to have regions inside the cells that are nonstructured—at least transiently nonstructured—and then you can move things from outside to inside or inside to outside.

I think this is the way nature does it. The image that comes to mind is a caterpillar or something where you have something propagating along it. This is one way that nature can move substances. I think it occurs not only in plant cells but also in animal cells. I give some examples in *Cells, Gels, and the Engines of Life*. It was published some years ago, and a few things have changed. I think it's still mostly accurate now, with the exception of a few considerations that are slightly different based on what we found since then. But I think this is how nutrients can move from one place to another place.

JOHN: The classical explanation for nutrient transport across the membranes is that the cell membrane is impervious and that there are channels and pumps and transport systems that move nutrients back and forth—that we can only move individual ions across cell membranes. And yet, particularly in agriculture, there are some pretty clear cases where the evidence doesn't fit the theory. The one that I refer to all the time is the absorption of systemic insecticides by plant roots, where we're applying systemic compounds that are absorbed in their entirety by the root system and then transported throughout the entire plant. You've proposed that there might be some alternate mechanisms for nutrient absorption by cells. Can you describe that for us?

JERRY: I think it starts with the fundamental assumptions that you mentioned: that the membrane is impermeable and that pumps and channels are what govern the behavior of cells: what comes in, what goes out. I think that concept is just dead wrong. I know I'm in the minority, but let me explain.

First of all, most cells—not all but most—have negative charge inside them, net negative charge. If you stick an electrode inside the cell and measure the electrical potential difference between the inside and the outside, typically in animal cells, it'll be −50 to −100 millivolts. In plant cells it's higher—typically maybe −100 to −200 millivolts. But the inside of the cell is negative. This is reported by many people for many years.[10] I think there's no question about it.

The only question is how this occurs. What's the reason for it? Jens Christian Skou, Paul Boyer, and John E. Walker got a Nobel Prize in 1997 for suggesting a sodium pump. This pump lies in the membrane, and the sodium gets pumped out. If you're pumping out positive charges—this is a simplification of the theory—but basically, if you pump out positive charge, then the cell remains negative instead of neutral. The pump keeps going, and therefore, the cell remains negative. It's an active, dynamic kind of thing.

A different explanation is that the EZ water inside the cell is negatively charged. This is the reason why the inside of the cell has a negative charge.

I'm not the only one nor the first to question the existence of pumps and channels. Gilbert Ling was one of three people chosen from all over China to come to the U.S. and study after World War II. Many people think he deserves at least one Nobel Prize and maybe more, but he's been passed over so far. Gilbert did some experiments to check this idea that the membrane contained pumps. He poisoned the cell in many different ways. It was, you might say, "overkill." If you pull the plug—if you metabolically poison the cell—you're basically pulling the plug on the pump, and therefore, the effect of pumping should disappear right away. But it didn't. He could measure what's been attributed to pumping; that is, the electrical potential was maintained for many, many hours after that and then finally disappeared. He computed this potential, and with the most generous of assumptions given to the other side, the cell needed thirty times the amount of energy that it could possibly, conceivably muster to achieve this. He said, "It doesn't work. The evidence doesn't fit." Nobody has challenged this since then.

---

10 H. Lodish, A. Berk, S. L. Zipursky, et al., "Intracellular Ion Environment and Membrane Electric Potential," in *Molecular Cell Biology* (New York: W. H. Freeman, 2000), section 15.4, https://www.ncbi.nlm.nih.gov/books/NBK21627/.

Gilbert raised a number of other considerations that have become increasingly significant. The number of pumps that are now suggested to lie in the membrane is something like 2,500 or so. The number keeps increasing because in order to explain the separation of various ions or other substances, the trend is to suggest a pump. Everything got its own pump, and typically its own channel as well, until Gilbert said that there isn't enough room in the membrane to accommodate all of these pumps, let alone enough energy to power them, because the sodium pump itself—even under the best of conditions, he computed—uses 30 or 40 percent of all the energy that the cell can muster. At that time, the thought was that there was only one pump—the sodium pump. If you think that pump is using 30 to 40 percent of the cell's energy, and now we're talking about thousands of pumps, it just doesn't make sense.

Gilbert's evidence suggests that there's no such thing as a pump. It doesn't exist in the membrane. As for the channels, there are other arguments, and I discuss these in *Cells, Gels, and the Engines of Life*. Researchers now say that there are more than a thousand different channels for different substances—different ions and such. There are big ones and smaller ones, down to very small ones, even so-called aquaporin channels for water. If you imagined the lineup of all of these channels in the membrane—one for the biggest substance, the next biggest, the next biggest, and so on, in order—the logic is that each one will allow only one particular substance to flow through and will exclude all the others.

So imagine—let's take the biggest substance, like a basketball—and imagine a channel that will let the basketball through. When the channel opens and lets the basketball through, that channel must be able to exclude the softball, volleyball, golf ball, and all the others, right? How is it possible that the channel for the basketball, when it opens, can exclude all the smaller balls? You would think that all the smaller ones would sail right through.

Logically, this becomes untenable because it's hard to think of a mechanism where a certain channel will pass its basketball through but can exclude all the smaller ones. There's a logical inconsistency in the idea of channels. Both the ideas of channels and pumps—there's something wrong with those concepts.

So we go back to the issue of how the cell gets its electrical potential. I'm suggesting that the negatively charged water that's inside the cell accounts for this. The idea has been that the cell membrane is impermeable to everything and that the only way things get in and out is through these channels. But if the channel doesn't exist, then what's going on? I think the answer is that the cell membrane is not impermeable to substances.

There are several pieces of evidence for this. The one that I like is where you take a cell and you pass an electrical current through it. This technique is called "electroporation"—once you have the electric currents, the holes in the membrane are large enough that genomic material can pass through. We're talking about really big molecules that can pass through. The evidence for this is that if you take new genetic information and put it outside the cell, it gets incorporated, and the cell expresses those genes. This means that the new genetic material must be getting from outside to inside the cell, and it must be getting from the outside to the inside through those holes that are blasted in the cell by the electrical pulse.

This can occur once you pass the electrical current. You can put the genomic material outside a day later and still get them. It must mean that these giant holes in the membrane can last for a day or so. Then you ask the question, "Well, something is wrong here because if you have a big hole in the cell, how come all the ions would go out or in and kill the cell, but the cell doesn't die?" It must mean that the membrane is not impermeable to all of these substances. And the reason the cell hangs together, even without a membrane that excludes everything, is that it's a gel. If you handle a gel, the gel hangs together. It doesn't need a membrane. And by the way, gels also came with negative electrical potentials—same as the cell—but they have no membrane. I would conclude by saying that the idea that the cell membrane is impermeable to all of this stuff is questionable.

Another experiment to support that is if you take a cell, like a muscle cell or a nerve cell—it's been done with both—and slice it like you slice a sausage, the cell doesn't die, nor does it generate a new membrane. When you slice, basically, you're cutting through the membrane, and you're exposing the inside of the cell to the outside of the cell, but the cell doesn't die, nor does it generate new membrane. This must mean that the concept that you must have an impermeable membrane around the cell is not correct. It is very fundamental. I think if one looks objectively at the evidence instead of merely accepting what textbooks like to say, one would come to that same conclusion.

Substances can get absorbed by the cell because there is no impermeable barrier. I know this is a highly radical point of view, but I think if you look objectively at the evidence, you can understand that this might be valid.

JOHN: Jerry, that's an incredible answer. It's not a surprising answer, given some of the evidence that we have observed in the field. And yet it is still a bit mind-blowing, considering what the implications are and what they might mean.

JERRY: Yes. I'm sorry to indulge in philosophy too much, but in my long career, I've seen that simplicity is the essence of science, and when a concept grows so complicated, it's really hard to understand and follow, and it's full of assumptions. It's like an upside-down pyramid. It gets too big, too heavy on top, and it falls of its own weight. It's unbalanced. It's the science of nature, I should say. This is my point of view, and it's no different from the point of view that's been argued for so many years—that nature is simple. If a mechanism looks and gets to be progressively more complicated, chances are that there's something wrong with the foundational concept.

I think that's where a lot of biology is right now: extremely complicated, with scientists standing up at the podium and reveling in the complexity of the subject that they're studying, demonstrating how clever they are to be able to understand and cut through all the stuff and to understand these complex mechanisms that other people may have difficulty following, thinking that science is really not simple after all. It's complicated, they say, and they need to fill in all the gaps of this complex kind of construct.

I think in this sense that science has gone astray. Remember the principle of Occam's razor? It should be Newton's razor. Occam was a religious guy, and he was saying that when you're arguing for the existence or nonexistence of God, you have to keep the argument simple. If you have a complicated argument versus a simple one, the simpler one should prevail. It's more likely to be correct. And then it was Newton who took this idea and applied it to science. He said, "If you have two hypotheses to explain something, the simpler one is going to likely be the correct one." It's still attributed to Occam, but a few centuries later, when Newton came around, it was really Newton who brought the idea into science. Scientists have abided by that principle for so many years. But starting maybe a hundred years ago with physics, physics became complicated. Of course the question is whether it's correct or not. Basically, it became mathematics. It became abstract mathematics.

It's hard to understand the belief of some people that abstract mathematics are required for understanding nature. It might be true, but my own particular point of view is that I think simple mechanisms—cause and effect—are the ones that are likely to be the actual ones working in nature. A bit of philosophy thrown in there with my apologies! I guess I've been in science long enough that I get to have an opportunity to voice that kind of opinion. Perhaps not.

JOHN: I think it's a very valid opinion and something that you don't need to apologize for expressing because it's one that many people—at least many of the people that I communicate with—agree with very much.

I'm still trying to fully understand a permeable membrane instead of a nonpermeable membrane. I guess the simple question is, where is the limitation—or what is the limitation—of cellular absorption of compounds? What do they have the capacity to absorb or not absorb?

JERRY: Well, I can't answer that question. I don't deny the existence of a membrane. It's just that either the membrane may not fully cover the cell—everybody presumes that it's necessary—or the membrane itself may be permeable throughout. To understand how that applies to substances that are taken up by the plant, this is more complicated. I hesitate to use the word "complicated." I guess it's an area that I, for one, have not given a whole lot of thought to. But I can certainly speculate that if you have a region of the cell in which the water undergoes a transition from EZ water to ordinary water, if there's a region with ordinary water, it can certainly absorb substances from outside the cell because it should be within that water that substances can freely flow.

Once they're inside the cell, if this unstructured water region moves along—as we've demonstrated that it can—it propels itself along, then you can get substances that can go from outside to the inside, or in a similar way, from the inside to the outside. How exactly that works is something that we've not studied. I can't suggest in detail how that might work, but it is a really important subject, worthy of study. It's just that most people who study these kinds of things are convinced that you need to start with a foundation that involves an impermeable membrane and channels and pumps. If you start with that, and if it turns out it's not correct, as I was suggesting, then everything else will turn out to be extremely complicated—or in some cases, impossible. I hope some people begin studying that. It's extremely important.

What's the size of the molecules of some of these pesticides? Do you know?

JOHN: I don't know off the top of my head. Molecular weights range from the forties into several hundreds, if I'm not mistaken. It's not my area of expertise.

JERRY: Okay. Well, one thing is possible: that certain substances can destroy EZ water. They can turn it from EZ water into ordinary water. If a substance is on the outside, if such a substance on the outside breaches the water and starts to change that region of water from EZ water into ordinary water, then it's possible that that substance will penetrate the ordinary water and get dissolved in it. Then, if EZ forms again behind it, this is a way that this substance could be propelled from outside the cell toward the center of the cell or somewhere inside the cell. This is one possible mechanism for transport.

JOHN: Good. Thank you for that. It's something that I'm certainly interested in studying more about in the future.

Another area that I'm wondering about the implications of is the water crisis in agriculture today. What I mean is that we have soils that do not hold water very well. They've lost their aggregate structure and aggregate stability. Water doesn't infiltrate into these soils very well at all. Soils have also lost much of their water-holding capacity, leading to loss of organic matter and the items that I mentioned earlier—aggregate structure, etc. Are there any implications for EZ water in agricultural soils? How does understanding the fourth phase of water shape our thinking about water in the environment?

JERRY: Very good question. This is analogous to you and to me—to our drinking of water. Sometimes the water we drink will just pass through our system and come out and won't result in great hydration, and other waters will stick. Why? What's going on? I think the same thing is true of the soil. So the first point is that EZ water is the main water that fuels biological systems, but EZ water grows next to hydrophilic surfaces—not all hydrophilic surfaces but many of them. So the surface of, say, the soil, or what's in the soil, needs to be appropriate. If it's not appropriate, then the water that goes in won't stick or won't hydrate.

So the water holding capacity, what you mentioned, has to do with the surface of what's in the soil. If it's appropriate—or natural, I would say—then it holds plenty of water. If you put water either through rainfall or irrigation or however, the soil would hold that water properly. But if you change the soil in ways that don't let the soil material hold that water, then the soil won't be hydrated the way we expect it to be, and the water would just pass through or run down the slope or what have you.

This is the issue about the soil content. The surfaces of what the soil contains need to be the kinds of hydrophilic materials that hold the water. That water will be EZ water. They need to be the kinds of surfaces that actually build EZ water. And then, if the surfaces are right, then the EZ water will build on them. We studied many kinds of surfaces, and if you add water, some surfaces are really great about building EZ water. That's the water that is in biology, the water that's needed. Other surfaces are not so great. It has something to do with the molecular nature of the particular surface.

One hypothesis—or a speculation—is that the charges on the surface need to be alternating in some way and distributed in such a way as to form a template for the buildup of EZ water. If they're not, then it doesn't work. Biological substances are ideally structured to do that, when you think of proteins and nucleic acids and such inside the cell. In terms of the soil, the same principles should apply: that the material in the soil should be such that the surfaces of this material have charged distributions that are just right to build EZ water. If not, water will just run off. When you destroy the soil, you're destroying the water holding capacity, which—in terms of what we found—translates to the ability to build EZ water.

JOHN: That's fascinating. I'm really glad I asked that question because there is this idea within biological agriculture that when soils have a strong fungal and bacterial populations, there is some mechanism by which fungi—particularly mycorrhizal fungi—can access water and make it available to plant roots—water that plant roots themselves may not be able to access. While that may be the case, what you've just described is a mechanism by which they can actually attract water. And water will flow much better.

JERRY: Yes. I know that's a critical issue, being once upon a time a backyard gardener. But yes, the organic soils are, of course, good at that. We don't use any pesticides of any sort. These are just awful poisons for us. I think a lot of people are quite aware of this.

JOHN: Jerry, what is the question you wish I would have asked?

JERRY: I guess I had wished that you might ask me about the role of water in our health. But I know you're agriculturally oriented.

JOHN: Isn't it largely the same thing?

JERRY: Yes. In a way, it's the same thing.

JOHN: Yes, please! What is the roll of EZ water in our health?

JERRY: From our point of view, what we've learned is that water is absolutely central for health. It's not just a passive tub full of water in which the important molecules of life bathe. If you look in a biochemistry or a cell biology book, water is considered not quite so important. I'm not sure about plant biology—maybe it's different—I haven't read them. Those books describe it as just something to base the molecules on. In some books it's even hard to find the word "water" in the index because it's so secondary. The first chapter will always talk about water and its properties, and then the rest of the chapters will go on to fail to mention water at all.

But water is so central. When the proteins fold, that does the work of the cell. In a muscle cell, myosin and actin proteins fold in some way, and that gives us contraction. Then they unfold, and we get back to the relaxed state. But they're all surrounded by water, although biologists often forget that these molecules are based in water. The kind of water they're based in is EZ water. So EZ water surrounds—or jackets or blankets—each one of these proteins. And when the protein is about to fold, its normal environment is EZ water surrounding it. In fact, the EZ water and the protein are a single unit together. They're one and the same. You can't separate them.

If the cell is missing this kind of EZ water, or if it's deficient in it—or you might say it's dehydrated—it can't function properly. If you're dehydrated, you can't play another match of tennis and expect to win. You need to get rehydrated. The cells can't function properly unless they're properly hydrated. From my point of view, you need a full amount of EZ water in yourself for that cell to function properly. A good deal of health, according to this paradigm, centers on hydration, centers on, What can you do to build your hydration? I guess the same thing is true in the plants as well.

There are a whole bunch of simple things that you can do. One is the sauna. It can be dry or humid, but it's always hot. Associated with this heat is infrared energy. If you sit in a sauna, you're absorbing a whole lot of infrared energy, and this infrared energy is what builds EZ water. The reason—or one reason, I think—you feel good, reenergized, after you sit in the sauna, is that you're absorbing so much infrared energy that it builds EZ water inside your cells. And especially those cells that had been deficient in EZ water, they come up to complement. When you leave the sauna, you feel good. Either your brain feels stimulated and refreshed or your muscles don't ache anymore. This is just one easy thing you can do to improve your health.

Another one is what plants do naturally: earthing or grounding themselves because they're in the ground. There are a lot of biophysical and health studies now that show that if you ground yourself or earth yourself by either walking barefoot in the sand or on the grass, or if you immerse yourself in a mud bath or something, you're connecting yourself electrically to the earth.[11] It turns out—although I didn't know this myself until about a decade

---

11 James L. Oschman, Gaetan Chevalier, and Richard Brown, "The Effects of Grounding (Earthing) on Inflammation, the Immune Response, Wound Healing, and Prevention and Treatment of Chronic Inflammatory and Autoimmune Diseases," *Journal of Inflammation Research* 8 (March 2015): 83–96, https://doi.org/10.2147/JIR.S69656.

ago—that the earth itself has a net negative charge. I grew up as a student in electrical engineering. If one of my professors had told me that the earth has a negative charge—if you plug into the wall, you've got three prongs, and one is the ground, which connects to a vast ocean of negative charge—I would have said, "You're crazy," because it doesn't make any sense. This was first presented to me by a Russian guy who declared that my education was deficient because in Russia, every middle school student knows about this. In the US, almost nobody knows about it. Not that it's controversial. It's just that it's not taught.

The earth is full of negative charge. Negative charge is needed to convert water to EZ water. So if you connect yourself, if you "earth" yourself by walking barefoot or walking on beach barefoot, you're connecting yourself electrically to this vast sea of negative charge. If you're connecting yourself to this sea of electrons, then these electrons can come and be absorbed in your body to build EZ water. I think this is the most straightforward explanation of why you feel better when you ground yourself, because you're absorbing this charge, it builds EZ water, and your cells can now function better.

A third way that you can help yourself is by doing what you plant people know very well, and that is by juicing—green juicing. Take a freshly growing plant and squeeze the juice out of it and drink it. What you're doing is squeezing the juice from inside the plant cells. The water from inside the plant cells should be filled with EZ water. You're effectively drinking EZ water, which should very nicely rehydrate yourself. As I said, hydration is so important for function and health.

JOHN: You connected another dot for me. Today, many growers are starting to do field measurements with drones, with various cameras, capturing various wavelengths. This interesting phenomenon has shown up where soils are really biologically active—that have high levels of organic matter and very strong bacterial and fungal populations—have an extremely strong infrared signature. They are emitting infrared substantially more so than comparable soils that are less active.

JERRY: That's pretty interesting. When you build EZ water, there are lots of charges that are moving around from one point to another. It's basically charge movement that gives rise to electromagnetic radiation, especially in the infrared region. This could be an indication of charges moving around to build EZ water. It's not a necessary condition, but that would be one interpretation of what's going on. So that would fit very nicely. I hadn't been aware of that. That's interesting information.

Meeting *the* Demand *for* Nutrient-Dense Food

MATT KLEINHENZ

Will there be a time when more people are paying more attention to nutrient density? I think so, but we're not there yet. I think that interest already is strong within a small community of eaters or a community of buyers. Where there are these so-called beachheads, there can be growth.

I've been fortunate to have had conversations with growers in which I've learned so much. I've also been dismayed by some of the lack of understanding that folks sometimes exhibit about basic biology and the crops that they grow.

In my perspective, quality is the platform for repeat sales. One might be able to make the initial sale based on some other factor. But the repeat sale comes from quality; it's one enormously influential component of the overall value that the buyer sees in the product.

The single most common source of struggle I've seen over the years is the grower looking at their product only as a farmer—being unable to see it in a more comprehensive way—most especially as the buyer sees it.

Learn something new. Challenge yourself. If you've been hearing or watching or reading information coming from one primary source, check a second one. Get an alternate point of view. Do not become too comfortable with what one knows to be "true." Question healthily.

HOW MANY consumers make their food purchasing decisions based purely on price? What about flavor? Are any deciding to purchase one type of food over another based on nutrient density? For those who are, how can growers market their products to meet that demand and to receive a higher price for higher quality?

These are all questions that Dr. Matt Kleinhenz has been researching. Matt is a professor of horticulture and crop science at Ohio State University, and he serves as a state extension specialist. He earned his PhD at the University of Wisconsin and is an active member in the American and International Societies for Horticultural Science.

Matt is very interested in understanding the dynamics of produce quality and how these factors connect with buyers in the marketplace.

JOHN: Matt, you've done a lot of interesting work over your decades at Ohio State University and before. What are some of the memorable moments that have brought you to where you are today in the various projects you're working on?

MATT: Some of the most memorable moments specific to my career involve interactions with growers and other members of the industry who have brought me a question or an issue that they wanted some input on. After lengthy discussions or experimentation, or reaching out to other experts, we often discovered a partial solution, or at least a way forward to reach a solution. The questions often focused on how to produce a better-quality product—mainly to satisfy their customers but also, perhaps, for a larger aim.

JOHN: When growers are expressing the desire to produce a higher-quality product, is that demand often coming from buyers? Is that demand coming from perhaps an altruistic sense that the grower has of producing healthier food? What drives the desire to produce higher quality?

MATT: In my experience, John, it's been both. Sometimes the grower has their own view—their own ethics, their own aspirations. They simply see themselves as being part of a much larger picture—to enhance people's health and well-being through the quality of the material that they supply.

In other cases, it's a little less altruistic. It has a business sense behind it, and there may have even been complaints about the quality of the product or questions about how it could be improved. They're responding to a real business need.

JOHN: When growers have a desire to improve quality, what defines quality in their minds? There are some variations from crop to crop, but in general, what are growers seeking to produce?

MATT: That's a great question, and I've been privileged to try to address that question in different settings through the years—in one-on-one conversations but also in group settings, maybe even as a person who was asked to comment to a select committee. The answers have been different. We typically start by establishing what quality is, or how it's defined, in the eyes of that particular person or group. It has physical components, it has chemical components, it has biological components. These various components or properties essentially define or describe the whole.

They may be focused on one particular property. It might be the shape, the size, the texture, the aroma, the flavor, the color, what have you. They may be looking at one particular property or component, and necessarily so, but it's difficult—perhaps even unwise—to separate it from the whole of the product, unless there's a very specific application in mind.

I would say that there is rarely one particular component or property that is common to a grower group or an industry group. It ranges. But there are a very large number of properties that we look at in our fruit and vegetable crops.

JOHN: I've recently been a part of a number of conversations where growers are expressing the desire to improve nutritional value. Sometimes they use the language of "nutrient density" to talk about the improved nutrient density of food. My observations have been that, for the most part, consumers and buyers at this stage aren't really having that conversation—at least not on a large scale. An appropriate analog that they might be looking for would be flavor and aroma. But what have you observed? Do you think there is the potential in the future to have a major market demand for "nutrient-dense" foods?

MATT: Perhaps. First of all, nutritional value, or nutrient density, is a significant passion of mine, and I know it's a significant passion for other investigators that I work with. Some of them are not in the so-called public light as an extension person, or as a grower/adviser, but they are working tirelessly and quite impressively toward a similar goal, but within their lane, regarding understanding what nutritional value is and how it can be enhanced. It is a big passion of mine.

To the second part of the question, I think we need to be very specific when we talk about nutritional value and nutritional density. It's a complex topic, and we don't wave our hands and say that just because it's complex, that we can't understand it or that we shouldn't approach it. No, quite the opposite. We should stay on task. We should continue to focus on how a farm can become an instrument of public health, producing a more nutritious product, a more nutrient-dense product, a more flavorful product—a product that is more desirable to eat simply because of its chemical, biological, or physical properties. These are all parts of the process of enabling the food that we offer folks to play a larger role in maintaining or enhancing their quality of life, especially through their health status.

To the final point: Will there be a market? Will there be a time when more people are paying more attention to this aspect? I think so, but we're not there yet. I think that interest already is strong within a small community of eaters or a community of buyers. Where there are these so-called beachheads, there can be growth. Here in 2018, one-half of one percent of people might make decisions around nutritional value. Some year down the road, we might be able to say it's 5 percent or 10 percent. In that increase, there will be so many opportunities for enterprising growers to be a part of that process; it will become that much more noticeable.

JOHN: As I think about the question of nutrient density, it seems to me that there are perhaps a couple of different frameworks that people are using to think about this. On one hand, there is a conversation about food biofortification—growing specific plants that have higher concentrations of vitamins or minerals that are needed by a specific portion of the population, and growing biofortified foods can really help supply those nutritional requirements.

Then there is a second conversation happening around foods and crops that have inherent nutrient density by default, such as blueberries or kale.

And then there is a bit of a third conversation happening among people who believe that we have lost substantial nutrient value in all crops, generally across the board, over the last forty, fifty, or eighty years—whatever timeframe. Here in North America, in particular, I've seen this conversation come up quite regularly.

What is your perspective, particularly on this latter piece: Is there evidence to support that we have lost a lot of nutrient density in our food supply?

MATT: It's possible. As a matter of fact, I had a very lengthy conversation on this very topic just two days ago with another grower and members of his farm team. They asked the same question: Is there evidence that they've lost nutrient value over time, particularly within their fruit and vegetable crops? My area of focus is vegetables, but I'm aware that the picture is much larger and involves everything that we eat.

I know that there have been reports suggesting that that is the case—that the nutrient density of the US food supply, or perhaps of the global food supply, has declined over a number of decades. I have no reason to question those reports at this point in time, but I also have no reason to go out and shout from the rooftops that the sky is falling. I think the devil is in the details. As an academic and an investigator, and also hopefully as an educator and as a consultant/adviser, I have to be very clear on the source of the information, how the information was developed and how it's being communicated and what it could mean.

I look at the reports and the data, and I determine whether there might be understandable reasons for it. I'm not justifying it as a trend. Look at what has happened over the last fifty to eighty years in our food supply and the way we provide it, and what people consume and what they expect of it. It can't be seen as a separate phenomenon from how we've taken on our day-to-day lives and what that means to individual growers. Growers respond to what the market asks of them. They are businesspeople par excellence most of the time.

One of the forces that they face is demand. What does the market demand? If the market has not demanded nutritional value and does not reward growers financially for it but instead rewards them for abundance or consistency, or for other attributes, that is what they're going to provide. Who should fault them?

JOHN: I absolutely agree that growers are responding to the market not just because of buyer demand but, perhaps, also because of the mantra that has been adopted by the agriculture industry as a whole for the need to "feed the world." That has been a kind of a subconscious driving force. Of course, we have the economic incentives, which are a major incentive for growers to produce large volumes and abundance—perhaps at the expense of quality and perhaps not.

MATT: This is maybe a section where I push back just a tiny bit. I'd say that in an individual grower's decision on how they approach the management of their farm and their crops—of course, I can't climb inside somebody's head, but what I've observed over the years is that they have their farm and their own personal welfare in mind first and the health of the industry second. They have to have a hierarchy; they have to have priorities.

We don't necessarily point a finger at the industry and say, "This is where the industry has led us." No, these are day-in, day-out / season-in, season-out decisions that a grower makes regarding what they grow, who they sell it to, what price they receive, and what is required to make it all work—little by little, farm by farm, region by region, industry by industry. The cabbage industry or the lettuce industry—various industries—they are not monolithic. There's quite a diversity within each. But if you look at them farm by farm, area by area, generational transfer over time, people have chosen to respond to the opportunities that, so far, on the whole, do not reward producing higher nutritional value simply because, I would say, the math is not in their favor.

There just aren't enough people that will pay more for the same product, if it's claimed to be more nutritionally dense than the product that's not—because they're responding to their pocketbook, right? If I go to the farmers' market and I have a head of lettuce that looks in every way identical to the one that my neighbor and competitor has on display, and mine is twice the cost, but I try to make a claim that it is more nutritionally dense, how many customers will walk away without buying the product?

Moreover, what evidence will I provide that my claim is true? And on what will I base it? We talk about nutritional value as if it is one thing, but it is most certainly not one thing. There are many components of nutritional value. But most of the time, the conversations speak of it as if it's eye color; it's not.

It's a very complex attribute of a food item, and to understand it fully, I think one needs to really study it and be prepared to adopt, at least temporarily, some reductionist thinking, and then to reassemble all those little pieces that come from that reductionist thinking into something that's more holistic.

JOHN: I absolutely agree with you. I think we are saying the same things. I believe this entire conversation has been one area of confusion, particularly within the organic industry. It seems that every few years an industry report gets published saying that organic foods have been discovered to not be any more nutritious than conventional foods. Then, a short while later, there's a rebuttal from the organic industry that, in fact, organic foods are supposedly more nutritious than conventional foods. I've seen this cycle repeat itself multiple times.

And yet, when you stop to think about it, organic certification has no parameters for nutritional quality. It does absolutely nothing to certify the quality of the end product, in terms of nutritional value or otherwise. It is purely a process certification that says a certain process was used that hopefully should substantially reduce the number of toxins and pesticides that might be in the food supply. It makes no reference to quality or to nutritional value. I think there's this conversation happening around nutritional value that really has no basis in the regulations or in the certification process.

I'd like to switch gears for just a bit, Matt. In all the work that you've done with growers, working with various crops, what has been the one thing that has puzzled you for a long time?

MATT: I've been fortunate to have had conversations with people in which I've learned so much. I've also been dismayed by some of the lack of understanding that folks sometimes exhibit about basic biology and the crops that they grow.

How could one practice medicine without knowing anatomy? How would one practice law without knowing basic precepts? How would one practice as a plumber and not know "lefty-loosey, righty-tighty"? How would one be as effective as they possibly could be at their chosen profession without a firm grasp of some of the basics? I've noticed the capacity for people to produce a lot of material because they follow a process. But they don't know the finer points of the process nor the finer points of the organism that's the central part of it. It's almost as if they're succeeding thanks to the process that was laid out for them by others, not because they have an in-depth understanding of the crop. Occasionally, that lack of understanding leads them to a problem that befuddles them. It holds them back and may even undercut their income. Of course, they understandably get confused and frustrated. In some cases, the answer is right before them—if only they had understood a little bit more about their crop.

I guess the one thing that has struck me is the lack of understanding of certain basics around each of the crops that folks grow.

JOHN: Matt, this is an incredibly important piece, and I had no idea how you would answer this question. I've been chuckling to myself because this is something that we have observed as well, quite frequently, in our consulting work.

In areas where crop production has become more intense and more competitive—where core crops are managed very intensely and margins become a lot tighter and the risk becomes a lot higher—it is almost invariably the growers who don't have the foundational knowledge that end up not surviving, for whatever reason. It might take a few years, it might take a decade, but invariably, it does seem that these are the growers who don't survive in the long term.

Following up to that, from your perspective and observation, what are the important pieces that you would recommend growers to study and to become really familiar with?

MATT: Great question, and I'd hasten to add that the understanding that I spoke about can come from many different places. There are different ways of coming to understand something as complex as lettuce growth or tomato growth.

I would include in that, obviously, what they're doing on a day-to-day basis—the business of farming, the process of farming. Take notes and observe. If a camera is an option, take pictures and record. In down times, when there are any, study those notes; look at those pictures. Think about what was going on at the time. Install ways of collecting more information.

There are very simple devices to record temperature. One can measure rainfall and so forth. There are flow meters on irrigation. There are steps that can be taken that will describe the system. In the process, growers haven't really changed anything about their farming practice; they simply know more about it. Soil testing would be another obvious example. Because if at some point, they need to turn to somebody for input, maybe somebody like you or me, we value that information. If I had more input, if I had more data, if I had more information, I could be more effective at helping growers.

Just observe, take notes, record as much as is reasonable, and be strategic about it. The second part would be to read, listen, and watch credible videos. Have a healthy skepticism about what you're seeing and reading. Don't accept everything at face value, but don't be so skeptical that you accept nothing. Be an informed consumer of information. As you deal with the confusion that can come with that healthy skepticism, there still will be growth.

In the long run, the quality information rises to the top. You can rely on it; it can be put to better use. Employ as many ways of knowing as you can, consistently, and just begin with the simple ones.

JOHN: A quote from William Albrecht comes to mind. In one of his articles, he wrote "Read books and study nature, and when the two don't agree, throw out the books." Of course, there's an element of humor in that, but I think that quote perhaps encapsulates a bit of what you're describing. In order to be able to throw out the books, you need to have high-quality observations. Which goes to the point that you were making—that observation is absolutely key in being able to understand what's going on.

MATT: Absolutely. Through those observations, one might be able to write their own book, which maybe somebody later will throw away. Or maybe not. Maybe that book will remain quite viable and informative for a very long time. I don't want to say there's a lifetime to all information so much as there is a lifetime to how it is applied in certain ways. What we see in the evolution of understanding people's businesses is that they change what information they rely on or seek.

As they become more proficient at certain areas of their business, they no longer need certain types of input as dramatically as they may have needed them some years before. Simultaneously, they're seeking out other types of inputs. On the question of nutritional value, they could begin with "What is nutritional value anyway? How would a dietitian/nutritionist, a health professional, define it?" I want to go to the primary source and ask what they do with it.

There are people—professionals of one stripe or another—who have written some very impressive resources on the nutritional value of various food items. Learning from those primary sources can be enormously important.

Ways of knowing about the farm—what to do with one's soil, what to do with whatever aspect of the farming operation—if one looks for input and doesn't find it, that's a clue that maybe it's not been investigated; maybe it's not been written about. They are there at the cutting edge of understanding it.

JOHN: One characteristic we've observed in growers who are exceptionally successful is that these most successful growers seek to understand how they make money. The key to having a successful, profitable crop is often some quality component—whether fruit size or fruit shape or a specific firmness. There are often very specific fruit characteristics or vegetable characteristics that are desired by the market.

One of the things that we've observed is that the most successful growers seek to understand that specific characteristic and how to manage it. They don't delegate that piece to an agronomist or a consultant or an adviser. If the key to having a profitable tomato crop is having a high soluble solids content, what are the specific components and factors that contribute to achieving a high percentage of soluble solids? They focus in on that piece specifically. They want to make sure that they personally understand that very deeply; they don't delegate that knowledge.

I think this speaks to this entire conversation: What are the important pieces for us to learn about as growers? I think really understanding the pieces that contribute to profitability can be a very valuable place to begin.

MATT: NO question. I agree with you completely. They don't delegate in part because, consciously or subconsciously, they recognize that quality is the target. In my perspective, quality is the platform for repeat sales. One might be able to make the initial sale based on some other factor. But the repeat sale comes from quality; it's one enormously influential component of the overall value that the buyer sees in the product. Of course, there's a price-quality interaction, but one never really wants to be racing for the bottom price, do they? In that scenario, the importance of quality is diminishing. As the price rises, the value of quality in that calculation increases in the buyer's mind.

My argument has been, "You want a better price? Raise the quality." Improve the value of the transaction that you're trying to consummate with the buyer by offering them something in return that is going to prompt them to give you more money for the product. If you go to the produce auction and you see what's being bought and what's not, that is immediate feedback on what sells and what does not sell, correct? There's no more dramatic demonstration than at the farmers' market on Saturday, or at the produce auction, or wherever one can see another person's product and what is being purchased and at what price.

To be a student of that transaction, I think, is one way of improving one's operation: understand why yours was purchased or not, and understand why it got the price that it did, and how it could possibly go higher. In as much as quality is a force in that, understand what the buyers are responding to. It could be the color; it could be the shape; it could be the consistency. Whatever it is, remember the physical, chemical, and biological components of it. If it happens to be nutritional value, or the holistic expectation that the buyer has about the product enhancing their diet in some way, study it and understand the origin.

I know that Brix, or total soluble solids, is on many people's minds. But Brix is not a nutritional component. It has little bearing on nutrition—specifically, nutrition as people who study nutrition for a living define nutrition. Brix has other implications, but they rarely include nutrition. A grower rarely delegates what's most important because it tends to be something they want to have more control over, something they want to be more informed about. In the case of quality, it's often the target of what they're trying to produce.

JOHN: MATT, what do you believe to be true about agriculture today that many other people don't believe to be true?

MATT: I hear so many different perspectives on agriculture. I hear that it's going to heck in a handbasket. I hear that it's enjoying a complete revitalization, and everything in between. I hesitate to point to one area and say that this is something that people don't believe to be true. But I would say that there are areas or topics that I hear about less than other areas.

One of those that I hear less about is what role the grower plays in our culture, in our society, in our daily lives. They're part of a fabric that is ensuring our access, as consumers, to healthy, nutritious, safe, and enjoyable-to-eat, pleasing-to-eat food day-in, day-out / meal-in, meal-out. They're kind of like this background group of folks that many in the population take for granted. That's unfortunate. But there are probably a number of reasons for that.

Why do I believe this to be true? They're often underappreciated, under recognized, but they keep on going, just like we hope that they would, and as they have for so long.

JOHN: Obviously, you've thought about this a little bit. If you believe that growers are underappreciated, which I absolutely agree with, have you given any thought to how that might be remedied?

MATT: Not to share an unpopular idea, but I think they need to ask more of themselves. I also think that they could benefit from being a bit more assertive, professionally, about what they do—how and why and what role they play. Some people will become activists, some people will become involved in grower organizations, some people will keep it one-on-one and simply have excellent conversations with their buyers on Saturday mornings or whenever they happen to encounter them—all of which are necessary.

But on the whole, I would encourage folks to use the opportunity to display their understanding of the farm's role and of food's role in that person's life. Because too many of our eaters more or less just eat to avoid being hungry, right? Which is entirely fine. It's a choice. Others, though, look to food for other types of return on investment; it's an enjoyable experience. When a grower encounters a person or a market that might be wired that way, that might be thinking of food differently—having higher expectations of it, ideally—I'd like the grower to be able to step up and say, "Yeah, this is what we do and how and why, and we have data."

Growers should have a real foundational argument for whatever assertion they would like to make. But a lot of growers, understandably, are on the farm doing their thing; being out and about and mingling with the masses is not necessarily their forte or interest. But there are other ways of having an impact and playing a role.

JOHN: I completely agree with what you've described. The growers who have been able to really make an impact with consumers and with eaters are those who communicate why. They communicate why they make the choices they do. There certainly is an element of describing what they do and how they do what they do. But I think the most important piece, which resonates most deeply with eaters, is describing why you need to make these various decisions and these various choices that the consumer may or may not necessarily agree with. When you describe the situation of what's happening and why you need to make these choices, their appreciation for the challenges and the difficulties and the opportunities in agriculture completely changes.

MATT: We want to be careful here. We want to elevate as much as possible the position of the grower within the whole spectrum of our society and our culture, and we want them to be successful. Thinking more about your question, the single most common source of struggle I've seen over the years is the grower looking at their product only as a farmer—being unable to see it in a more comprehensive way—most especially as the buyer sees it. Attitudes like, "Well, if I eat it, everyone else should be able to eat it too."

When a grower is able to see the product from completely the other side of the table, as they—as much as possible—shed their grower attitude temporarily and see the product as if they're buying it, for how it's going to be used by the buyer, then they're on a really exciting path. Then they can look differently at their own farm and possibly be able to exploit market opportunities that they didn't see before.

That's the single most consistent aspect of a vegetable farm now, especially a vegetable farm that might be selling directly to consumers, but even those who grow for processors.

JOHN: That's a fascinating observation.

MATT: They need to see it as the buyer does, and unequivocally so—without reservation, not kicking and screaming. They need to welcome the opportunity to see the product as the buyer does. You will return to being a farmer—no worries! But when you return to being a farmer, hopefully you carry that experience of seeing the product differently with you. All of a sudden, for some, it will be like, "Actually, I'm not producing kale; I'm producing food that someone's going to serve at a family function, and it's got to be just so."

We do these exercises with students in the class where I hand them a tomato, or I hand them a potato, and I say, "Tell me what you see." And everyone looks at me funny. You'd be amazed by the kinds of words that are used to describe ordinary products. But then we're on a path toward understanding what that tomato is. If you go through that same exercise, for example, like I have, with students in the dietetics nutrition arena versus the students in the agriculture arena, it's amazing how they look at the same thing and use different words to describe it.

For a grower to understand how others see the product is indispensable.

JOHN: Matt, this is such a fascinating observation. I'm smiling to myself because I'm recalling a farm that a colleague of mine worked on a number of years ago. It was a large commercial farm growing salad greens, in particular lettuce. They had a couple of different lettuce fields that were affected by a major hailstorm within about eight or ten days of harvest.

When they visited the first field to inspect and evaluate the damage, my friend asked the grower, "When you think of this field of lettuce, what does it mean to you? What do you think of, and why are you growing lettuce?"

The grower's response was, "Well, I'm growing lettuce. My goal is to have eight hundred cases per acre. When I think of it in my mind's eye, I see it as eight hundred cases packed in the cooler, ready to ship out, and I want to get a good price so that I can be profitable and keep growing this crop again in the future."

They went to the second field, and the second field had very similar damage—almost the exact same amount of hail damage. And my friend told the grower, "I want you to think about this field of lettuce not in terms of profitability—in cases of lettuce in the cooler. Instead I want you to imagine a young family with a couple of young children sitting down to eat dinner, and they are having a salad that is made with your lettuce."

They were just having this conversation—and I know I'm probably going out on a limb here quite a bit—but there was this intriguing crop response. When they evaluated the two fields, with the same amount of hail damage and crops of the same variety that had been planted the same day, when they went to harvest the first field, they only had about a 20 percent harvest. In the second field, where the grower had a different mental image of the crop that he was producing, they had an 80 percent yield. There was a tremendously different response from the plants in the field.

I know this is pretty far out there. Did those plants respond to a different mental image of what they were being produced for? I don't know. I'm not going to suggest that they were. But there definitely was something majorly different in those two fields. That observation really caught my attention; I've puzzled over what was going on there. What was happening; what did we miss?

MATT: Right. I clearly don't claim understanding of much of anything. You went beyond vegetable production! But as you observe other aspects of many people's daily lives, in terms of what they buy—their transportation, their clothing, their music, their appliances, so on and so forth—I think a number of very successful companies recognize that—perhaps as with vehicles—it's not just transportation; it's the feel that one has in the vehicle or when looking at the vehicle. There's something else going on.

I'm not trying to draw parallels between those directly, but I do wonder whether a fresh look at what is actually being offered for sale on the part of the grower would help—if a fresh look, a buyer-informed look, would go a long way in helping them day in and day out.

People say, "I get that kind of feedback when I see what's left over at the stall, or when people come into the farm market and they pick what they pick or don't pick, and I see what's left over."

I tell them that I agree. But are they asking their customers in any other way what they actually came in for? It's possible that they took something away simply because they had to; they didn't have enough time to make another stop. But they won't be back. Or they're not overly enthusiastic about what they took away. They took it because they had to: it was a matter of timing. But they're calculating in the back of their mind, "I'm not going to be coming back. This will not be my first destination."

The grower has an internal control on that day, at that time, at that point of sale. But do they really have a more comprehensive understanding of what the buyer was looking for? Maybe not. That comes with some real sophisticated but focused attention to what the buyer is actually looking for. More of that would help.

JOHN: ARE you asking the deeper question, perhaps: What is the job that the buyer wants this food to do? Such as, does this food have a purpose—that it needs to look really good on the Thanksgiving dinner table, for example? Are you asking more about its purpose—its greater purpose other than just nourishment?

MATT: Certainly. But I'm advocating something perhaps more mundane, more day to day. What I'm advocating is for growers to be as informed as possible about what their buyers expect, as a minimum, and what they aspire to.

Because remember, buyers—especially those who are chefs, who are reselling the product, or produce managers at grocery stores—intermediaries, if you will, and even some individual consumers, they have aspirations too. We're talking about how growers can grow and adapt. So are other businesspeople and individuals: they're growing and adapting. We see evidence of this quite profoundly.

For the grower to be in step with that market, that individual, that group—or to even sidestep them, to make a parallel move and become more informed about another group—is only to their advantage. Wherever it leads, to be more informed about what the buyer is expecting, right in the here and now, and what the buyer would reward later, is only to the grower's advantage.

I'm part of a much larger team at Ohio State University that is involved with understanding or working to understand connections between diet and health, particularly within a defined community of eaters, namely cancer survivors. These are folks who have gone through their primary treatments and are now on the other side of that. Some subset of them is being provided access to freshly harvested fruits and vegetables.

They have an entirely different outlook on diet. Having gone through their diagnosis and treatment, they now have a different appreciation for and view on diet-health relationships. They have an entirely different view of what they will buy and what they will look for. Some of these folks, of course, are midlife and older. And they're not going through this scenario alone—they have families, they have friends. My point here is that with their revitalized appreciation for diet-health connections, and now a focus on what they eat, are growers being a partner in that process? Are we taking any steps to connect with people who have a new interest in food? Or are we going to leave it to chance?

Are we going to leave it to them to go into their local store to talk with the produce manager or their restaurant or whatever it might be? Is it going to be just chance, and the grower will get whatever comes their way? Or will growers be proactive in that process? Of course, when you're asked, "Can you grow a more nutritious product?" be able to say, "Yes, indeed I can. And it will have elevated levels of this, this, and this." Because the dietetic, nutrition, and oncology communities will have certain expectations of them. It isn't just like waving hands and saying it's more nutrient dense. No, there are specific bioactive compounds that matter.

Will the growers know about those? Will they be able to manage their crop accordingly? If we want to get to this higher plane, everyone needs to step up their game. Advisers, scientists, growers—we're all being asked to do more. But in that, there's greater opportunity as well, I think.

JOHN: From your perspective, where is the greatest opportunity for growers today? Where's the greatest opportunity in agriculture? Where's the greatest opportunity to move to that next plateau?

MATT: Growers should continue to do what they have done for many years very well, but do it better. We focus now on social implications of farming and farming systems and agriculture overall. We focus on the social component. That's very, very important—so-called food equity, food access, food availability, food security. Be a partner in the process to some level in enhancing the social component of what they do on their farm and be connected in some way.

Secondly, environmentally—yes, we ask a lot of growers. They're supposed to do this, this, and this, and meet these criteria. But growers should do their best to use the natural resources that are required to produce food well.

Be a good steward of the earth. There will probably be rewards for that. One can be certified organic or not, but be a good steward of the natural resources that are required to produce food.

Finally, be a steward of the farm, in that you're trying to run a business. You're trying to be successful at it. You're trying to generate an income. Take as many steps as possible to understand the market and to enhance the price you get. Be a student of it; explore it.

I think those are possibly the greatest needs, or the greatest opportunities.

JOHN: If you had to make one recommendation for growers to learn more about any of these various areas—if you had to recommend one book for people to read—what would that be?

MATT: Maybe not one book so much as one step: learn something new. Challenge yourself. If you've been hearing or watching or reading information coming from one primary source, check a second one. Get an alternate point of view. Do not become too comfortable with what one knows to be "true." Question healthily.

If someone is only reading the scientific literature, he or she should look at some other literature as well. If one had no exposure to the scientific literature, no exposure to research, no exposure to that kind of information, I think it's high time to start. If one has never picked up a book describing the physiology of tomatoes or kale or whatever, I think it would be a good time to do that. Challenge yourself to learn something new about your crop and/or your farm.

JOHN: Your earlier comment reminds me of a quote that I believe is attributed to Charlie Munger, where he describes the work required to have an opinion. He says that in order for you to be allowed to have an opinion, you need to be able to articulate the opposing point of view better than the opposition.

MATT: We see this in some high school speech and debate programs, where knowing the opposing point of view is essential, even though you're going to espouse your point of view. To see the issue in the round is very, very important. You're still free to make the choice to adopt one point of view or another, but being familiar with the opposite is essential.

At least as an investigator, as an academic, that's kind of part and parcel of our careers. Because as we see new data emerge, we have to revise our thinking on a topic. And academics are fairly adept at picking out folks who are operating with "old" thinking. It's just something you become somewhat adept at.

Some of that is okay; it's acceptable. It's a challenge to remain up to date, but it's necessary. It's the same for farming. If a grower is not aware of some of the newest trends in technology and what people respond to in terms of their food, like you said earlier, it's only a matter of time before that so-called ignorance will perhaps have undesirable consequences.

JOHN: SPEAKING of technology, what technology or ideas are you really excited about for the future of agriculture, that you believe hold a lot of untapped potential?

MATT: Given the types of operations I interact with, I would say there are two sectors that are related. One is grower-friendly technology that can be implemented on the farm in terms of monitoring the environment and understanding the connections between that environment and the cropping outcome. These are sensors—data loggers, miniature temperature sensors—the ones that will communicate with the grower through, for example, a cell phone or a computer if they have one. Those are interesting to me, matched up with more computing power. We have more computing power in a standard laptop than what went up on the Saturn rockets into space. It seems like, isn't that enough? But I think it's the analytical capabilities that we're beginning to implement that will matter to growers going forward.

I think there's a rather large untapped potential out there for growers who are connected to some types of databases. For example, how much does a grower know about the people within a fifty-mile radius of their farm, regarding any factor that would influence their food choice? How much do they truly know?

It's not about prying and snooping on the internet. It's about understanding data about the demographics of the people around them. Understanding those data more thoroughly may give them insights into how they could reach them. When they reach them, what would be valuable to them as somebody offering vegetable crops, for example?

Is there a new ethnic population coming into the area that would be looking for a certain type of vegetable that that grower could provide? Is there a new employer coming into the area? Is there an employer leaving the area? Any number of factors influence food choice and how people interact with their food, particularly vegetables. This would be information that they could employ if they happened to be active in the local or regional markets. Obviously, people who are shipping all over the country are tracking this anyway.

JOHN: Matt, we've covered a number of very intriguing topics. What is the one question you wish I would have asked? What's the piece that you would have loved to discuss in this conversation?

MATT: I'd like to return to the question of nutritional value and so-called nutrient density—a term that I and others have instead called "nutritional yield." It's a combination of the traditional yield as we measure it—pounds per acre times the nutrient value it contains—and the concentration of this product or that metabolite or another metabolite. That's the metric that we have been using in some of our studies to understand how we can move the needle on nutritional yield, recognizing that, for the moment, most growers are paid by the pound and not so much by the nutritional value.

We anticipate a time when the price as a proportion of the normal yield and nutritional value will change. In other words, when people value nutrition more highly. I'd like to ask growers what they're hearing. What are they seeing from their customers regarding the role of nutrition or quality, flavor, texture, and so on. I like to learn more from them to get a sense for whether or not my reading of the situation is accurate, in order to guide the work going forward.

JOHN: There are a couple different questions here. There's obviously the question of food quality. From perhaps a macro perspective of needing to grow better nutrition—or I should perhaps say more nutrition—for a growing world population, should we, perhaps, also begin measuring crop production in terms of protein content per acre or carbohydrate production per acre for what will actually be used as a food source? What are your thoughts?

MATT: Some of that is already being done. Perhaps not much is being made of it publicly. The composition or the makeup of the major components of the food supply in the US, and Europe and other places, is tracked. There are federal and state agencies that do that. It's mostly federal agencies. That's what they're tasked to do. We have some really capable people who are not only involved in creating those data but also in analyzing them. We can model, for example, what the impact would be if, as a population, we ate one less pound of meat per year, per capita. What could that mean in any number of ways? Or if certain parameters, certain properties, of the food supply went up by X percent, what could that mean from the standpoint of diet and health? We have some projections, and they're very intriguing.

In as much as a grower would try to build part of their business around that process, I am acknowledging the day-to-day realities that most farmers appear to face. It's a huge challenge to do much of anything other than producing the crop using the ways that are very familiar and reliable. To add on to that—"Wow, you're asking me to become an expert in nutrition?" No, just take a few steps forward in that direction to appreciate and understand where there might be a market opportunity for you to grow the business based on the quality of the crop and the markets that you're now trying to access.

As much as you can, shape the market rather than just respond to it. To whatever extent possible, create the desire; create the market. Don't just respond to it.

JOHN: How would I, as a grower, work to create the market? How can I shift the perspectives in the market?

MATT: You'd be as informed as possible about some of the topics we covered—about what people expect in the product. Be conversant in them so that when there's an opportunity, you can bring up the quality of the products that you're selling—the ones that can't be seen. Some of these nutritional parameters can't necessarily be seen, and some can be seen very well. Color is a prime example of that.

If you study the dietary recommendations, as drawn by the dietetics, nutrition, and human health professional communities—not the USDA alone—you see that color on the plate is enormously important because from color comes so many benefits. In the fruit and vegetable world, we obviously have a huge palette of color opportunities.

Understand what it might take to provide a deeper red or a brighter yellow or a more penetrating blue or purple and what that could mean nutritionally, and from the standpoint of the flavor profile—we can't ignore that—and be able to converse about that with the buyer. Understand where nutrition comes from. Nutrition is defined by the people who study it directly.

About the Author

JOHN KEMPF is an entrepreneur, a speaker, a podcast host, and a teacher. He grew up on a fruit-and-vegetable farm in Northeast Ohio, and he has developed nutrition management systems that regenerate plant and soil health quickly, resulting in strong economic returns for growers.

As the founder of Advancing Eco Agriculture, Crop Health Labs, Ozadia, and the Regenerative Agriculture Academy, he is passionate about the potential of well-managed agriculture ecosystems to reverse ecological degradation, bringing joy and profit back to producers. John's mission is to have these regenerative models of agriculture management become the mainstream globally by 2040.

Index

D

E

F

N

O

P

Q

R

T

U

V

W

X

Z

Made in the USA
San Bernardino, CA
26 June 2020

74426781R00166